CRYSTAL BALLS OF
THE 21ST CENTURY

CORRUPTION OF REAL MONEY (PART VI)

FIRST EDITION

MARCO CHU KWAN CHING

CRYSTAL BALLS
OF
THE 21ST CENTURY

FIRST EDITION

MARCO CHU KWAN CHING

Crystal Balls of the 21st Century
Copyright © 2020 by Marco Chu Kwan Ching

Cover Image © Ivaylo Sarayski | Dreamstime.com (Royalty- Free License)

Published by Marco Chu Kwan Ching

ISBN: 978-0-6486664-7-9

"To Mom and Dad"

- Marco

Acknowledgements

Where do I begin thanking all the people who helped to make this book possible? This book represents one of my most dedicated missions of my life.

I would like to express my gratitude to my parents, Angela Tsang and Tony Chu, for their encouragement. I would like to give special thanks to Mike Maloney who started me on the road to invest in precious metals; and to Peter Schiff for his unparalleled economic insights; Dr. Richard Buckminster Fuller for his world views; Richard Heinberg for his education on energy and ecological issues. I would like to thank Michele Berner for proofreading my book; I especially like to thank my grandparents, as my childhood with them is instrumental in bringing this book to fruition.

Contents

Introduction

Part I: ***The Predicaments of the 21ˢᵗ Century***

Part II: ***Crystal Ball of the 21ˢᵗ Century***

CRYSTAL BALLS

OF

THE 21ST CENTURY

"Our Children's children, who haven't even been born yet, are counting on you!"

Introduction

Why did I write this book?

In 1972, a team of experts at MIT undertook a project at the *Club of Rome* to understand the limitations of our world system and the constraints on human activities and population. Dennis Meadows, an American scientist and a team of 16 researchers produced the *World3* model - a computer simulation on the interaction between population, industrial growth, food production and limits in the ecosystems on Earth.[1] This model was used by one of the *Club of Rome*'s studies to produce a pioneer report that later became *The Limits to Growth (1972)*. It is the best environmental book of all time.[2]

The book, *The Limit of Growth (LTG)*, questions the assumption of whether economic growth can continue in the foreseeable future. It argues that our biosphere has a limit to absorb more human population growth, pollution and economic growth. The authors fed data for the world population growth, consumption trends, the abundance of various resources, food production and pollution into the *World3* model and came up with a reality that economic growth will peak sometime between 2010 and 2050, followed by a decline in food and population. This scenario has not been run once but many times since it was published. Even with more recent data and more advanced software, the simulation shows the results have been frighteningly similar.

State of the World

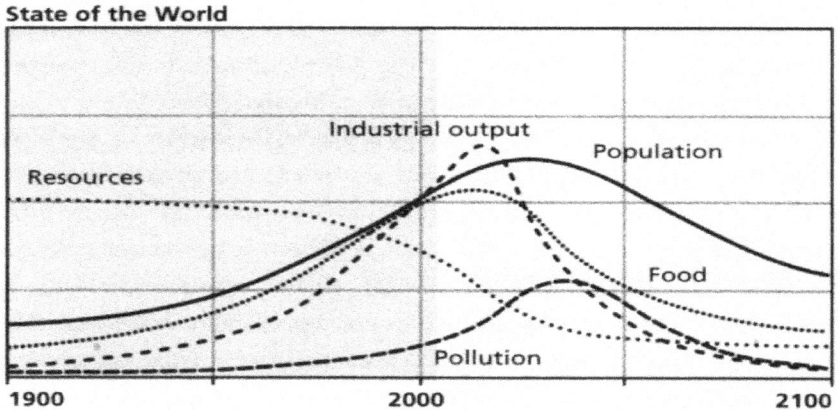

Figure 1: State of the World

Source: Limit to Growth

This book describes a possible future in the 21st century. I believe the world is heading towards a dire future with unsustainable trends in economy, energy, and environment. At the same time, humanity is also heading towards a technological singularity. I believe this singularity will be a turning point in human history. It is a double-edged sword that might help us to overcome the coming crisis and concurrently, create new ones. Either way, there is a lot to be done to create a future worth inheriting. Like the *grandfather of the future*, Richard Buckminster Fuller, once said: we are called to be architects of the future, not its victims. The information present in this book contains essential information for you to be one of the architects.

The Grandfather of the Future

Dr. Richard Buckminster Fuller (nicknamed Bucky) was an American author, architect, inventor and visionary, who believed all of life is a grand experiment by the Great Spirit. Fuller was awarded 28 U.S. patents and awarded nearly 50 honorary doctorates for his work in science.[3] He is famous for the geodesic dome and was considered one of the world's greatest geniuses. By 1980, there were 90,000 published references to Fuller and his work through print and electronic media, including the cover of Time magazine in 1964.[4]

Figure 2: Richard Buckminster Fuller

Source: Time Magazine

Bucky was one of the world's first futurists and global thinkers. Before his death on July 1, 1983, he predicted that there would be a technology that would dramatically change the world before the end of the decade (i.e., 1990). On August 6, 1991, ARPANET became the Internet and went live to the world.[5]

Bucky is convinced that provided it is properly used, technology could enable him to improve his environment to a point where machines could take over all work, and man would devote to learning more ways of *doing more and more with less and less.*[6]

Spaceship Earth was a term Bucky coined to express the need for mankind to use teamwork to fully utilize Earth's limited resources instead of competing with each other for it. Bucky hypothesizes our planet as a spaceship – one that can grow life and regenerate by itself, fueled by another spaceship – the Sun. We need to look after it for it to function properly. Unfortunately, humanity doesn't understand this mechanism. We do not have Earth's operation manual.[7]

Bucky is well ahead of his time regarding the problem our planet will face in the 21st century. He envisioned global problems, such as poverty, increasing population and the uneven distribution of resources around the world. As a result, Bucky invented the Dymaxion Map - the first world projection to show the continents on a flat surface without visible distortion.[8] He thinks that traditional world maps reinforce the elements that separate humanity and fail to highlight the patterns and relationships emerging from the ever-evolving and accelerating process of globalization. The meaning behind the Dymaxion Airocean World Map is to allow people to strategize solutions to global problems, matching human needs with resources.

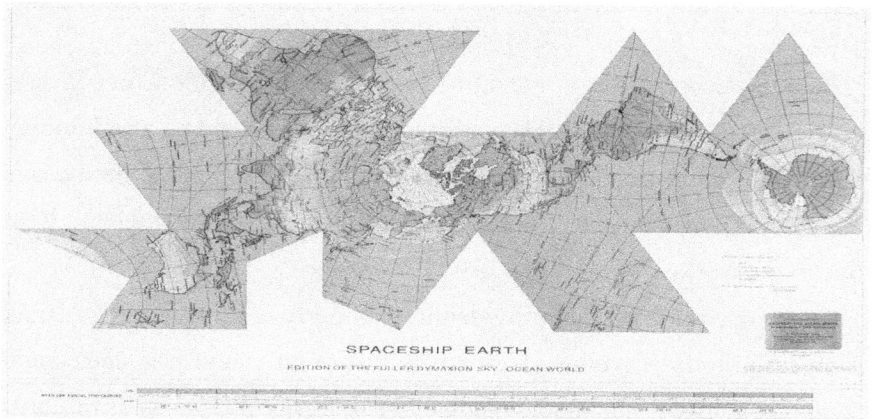

Figure 3: Dymaxion Airocean World Map
Source: Buckminster Fuller Institute

The Threat of Mass Unemployment

The Rise of the Robots: Technology and the Threat of Mass Unemployment is a book written by American futurist Martin Ford. I think anyone concerned with the future of work must read this book.[9]

In his book, Ford discusses the impact of accelerating change and artificial intelligence (AI) will have on the labor market. He predicted that this new wave of technological revolution will cause great social and economic disruption. Robots and other computer-assisted technologies would take over tasks previously performed by labor. Many of the jobs existing today would be outsourced to the developing world or replaced by AI.

The Coming Singularity

In April 19, 1965, *Electronics, Volume 38*, Gordon Moore, the co-founder of Intel Corporation, wrote that the future of integrated electronics would become the future of electronics itself.[10] The advantages of integration will bring a proliferation of electronics, pushing science into many different areas. With the unit cost of integrated electronics falling and the number of components per circuit rising, we are getting cheaper and more powerful computers. Moore's Law states that the number of transistors on a computer chip doubles every eighteen months.[11] Today, we have billions of transistors on an Intel i7 chips.[12] Because of the exponential growth in the powerful computer chip, the power of technology is accelerating exponentially.

In fact, by 2050, the power of technology will be a *billion* times more powerful. We have already experienced this type of growth if you look back in the 20th century. Comparing computers in the 60s to the iPhone at the beginning of the 21st century, you will realize that the pocket-size computer we all use today is indeed a million times *cheaper* but thousands of times more powerful in terms of price performance. Not only that, but you can literally assess all knowledge with a few keystrokes anytime, anywhere in the world. Ray Kurzweil, the author of *The Singularity is Near*, predicted that we wouldn't experience 100 years of progress in the 21st century — we are more likely to experience 20,000 years of progress (at today's rate) instead.[13] It is not just technology that is undergoing exponential growth, but the *exponential growth of technology* itself is growing exponentially. By 2050, it is predicted that machine intelligence will surpass human intelligence for the first time in human history – leading to a singularity.

The advance in computing technology will accelerate the technological progress of all other related technologies. One such technology is the reverse engineer of the human brain.[14] Today, it might seem a daunting task to scan the human brain with sufficient detail to be able to download. But, if you look back in history, the Human Genome Project (HGP) that began in the 90s was also formidable.[15] Discovering how the human brain works in the language of engineers will lead to transformative A.I. If successful, not only will AI be able to acquire human brain capabilities like emotional intelligence, but it will also illuminate new approaches to help the disabled who are blind, deaf, suffering from learning disabilities or age-related memory loss.

In the 1900s, a human's life expectancy was around 48 years old. Today, it is 78 years old.[16] In the coming decades, with the exponential growth in genetic, robotic and nanotechnology, it will become possible for us to re-program our biology away from cancer and heart disease to further extend our average lifespan. We will develop techniques to communicate and control our biology at nanoscale by reprogramming individual cells in our body[17], stimulate protein folding and interactions, 3D print new organs. We will be able to silence the gene responsible for our genetic diseases.[18]

It can be a very exciting future ahead, indeed.

We can all be architects of the future, not its victims. The outlook of the 21st century might look grim at present. Personally, I am confident that technological breakthrough can turn this around. After all, there is nothing in a caterpillar that tells you it's going to be a butterfly. For the first time in history, technology enables us to take care of everybody at a higher standard of living than any of us have ever known. With the unlimited creative ability of humanity and accelerating growth of technology, we will create an even more prosperous future that is worth inheriting.

Who is This Book For?

Crystal Balls of the 21st Century is a snapshot of a possible future in the 21st century. It's meant to be a book written for people who want to understand how we got to the point where we are today and where we will be heading in the decades ahead.

Humanity is facing unprecedented challenges in human history. We are heading towards a future where we have uncontrolled population growth, grave threats of climate change and rapidly depleting energy resources. We have a reality that our economy must grow, but our energy system and finite resource can no longer support that growth.

If you are picking up this book, you might fall into one or more of these categories.

1.   Futurists who want to know what is happening in the 21st century.

2.   Anyone who wants to understand the current trend in population growth on Earth.

3.   People who want to have a big picture of the current trend in population growth.

4.   People who want to understand the story of energy.

5.   People who care about the environment.

6.   People who want to understand the future of money.

7.   People who want to understand the history of the Information age.

8.   People who want to know how artificial intelligence works.

9.   People who want to know the future breakthrough in medical and genetic technology.

10.   People who want to know how nanotechnology will shape our future.

11.   People who want to know what the future of transport technology is going to be.

12.   People who want to know how industries will transform in this century.

13.   People who want to know about the 21st century science revolution.

14.   People who wonder if time travel is possible in this century.

15.   People who want to know about whether the future of humanity is in space.

16.   People who want to criticize my book.

Part One: The first part of the book is divided into four chapters. It aims to address the problems we face in the 21st century. The first chapter talks about the unsustainable trend of population growth and how it has overshot Earth's carrying capacity. The second chapter talks about the story of energy and why I think renewable energy is unlikely to replace fossil fuels in the decades ahead. The third chapter talks about the wrath of Mother Nature. The fourth chapter talks about monetary history and how the rise of the Global dollar standard changes the nature of the global economy.

Part Two: The second part of the book is about technology. It talks about how technology is going to shape our future and counter the problems described in Part One of the book. There are twelve chapters in this part. They are diverse topics in technology that will define our future.

Chapter five will talk about how technological growth becomes uncontrollable, irreversible, and unforeseeable to human civilization.

Chapter six will bring you back in history to look at the history of the information age, the birth of the computer, and the rise of Moore's Law. I will explicitly talk about how information technology has evolved from 2000 to 2020. This chapter aims to equip readers with background information about how we arrived at the point today regarding information technology.

Chapter seven will talk about the rise of artificial intelligence, how machine learning works, and the types of AI.

Chapter eight will discuss how cancer might be a disease of the past, how CRISPR will reform medicine of the 21st century. Apart from that, I will also discuss how our road to immortality and longevity might be attainable in the decades ahead.

Chapter nine will talk about how close we are in designing life.

Chapter ten is the most important chapter of all. It talks about how a new versatile material in nanotechnology will completely transform everything in this century.

Chapter eleven will talk about futuristic transportation technology that will redefine how we travel.

Chapter twelve will talk about Industry 4.0.

Chapter thirteen will talk about CERN, the particle accelerator, the discovery of the God Particle and the completion of the Standard Model.

Chapter fourteen will talk about how Quantum Technology will help us to create unimaginable technology.

Chapter fifteen will talk about plausible time travel technology and existing time machine.

Chapter fifteen will talk about how humans might become interplanetary species in the 21st century.

It is going to be an interesting journey into the future. So, buckle up, let our future begin.

Welcome to the Crystal Balls of the 21st Century

PART I

The Predicaments of the 21st Century

Chapter 1
Population Growth

Have you ever wondered if the population growth rate today is normal? Does Earth have a limit as to how much human population it can sustain? More importantly, have we reached that limit already?

Figure 1.1: Population Growth

Source: Author

At the beginning of the 19th century, the world population hit one billion for the first time in history. In just another century, the world population reached two billion. But, population growth did not pick up pace until we were officially half way in the 20th century. In the span

1

of just another three decades, the world population hit the three-billion mark. Since 1960, it has taken an average of just twelve years for the world population to double. As I am writing now in 2020, the world has already hit 7.8 billion in population.[1]

If we take one step back and look at the story of the world population, it took the world 11,804 years to reach one billion in population. But, since 1800, the world's population has increased 7.8 times.

It is a mind-boggling change. What happened in the 20th century?

Despite the Spanish flu epidemic and two world wars, the world population growth shows no signs of stopping.

The Picture of Population Growth

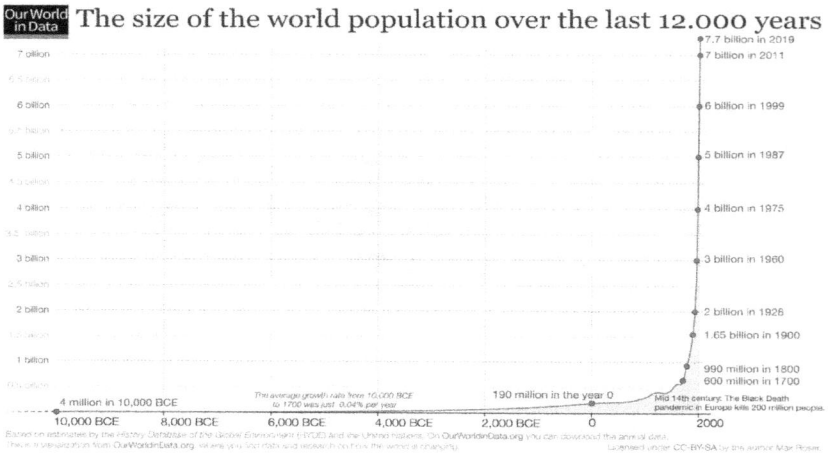

Figure 1.2 : The size of world population over the last 12,000 years

Source: Our World in Data

Year	Billion	Years takes to reach next Billion
1804	1	-
1928	2	124
1960	3	32
1975	4	15
1987	5	15
1999	6	12
2011	7	12
2020	7.8	12
??	9	??
??	10	??

Table 1.1: Population Growth

Population growth dictates our future. Unfortunately, right now, it is one of the most avoided subjects because a proposal to dramatically correct population growth is immoral and politically suicide. In reality, without truly addressing human population growth, all other solutions to climate change, energy challenges, shortage of freshwater, inequality, conservation of environment, quality of life, become irrelevant. I dare to say runaway population growth will offset all other efforts we try to solve the issues mentioned above.

But, hasn't our Earth always been supporting an ever-increasing population growth?

True. An ever-increasing population growth is normal, and we are accustomed to it.

However, many scientists do not believe this normal is sustainable. They believe that Earth has a maximum carrying capacity.[2] This carrying capacity depends on how fast nature can absorb our waste and generate new resources. If we use 1970 as a reference point, right now, we have overshot the Earth-carrying capacity by fifty percent.

What is overshoot?

How many Earths does it take to support humanity?

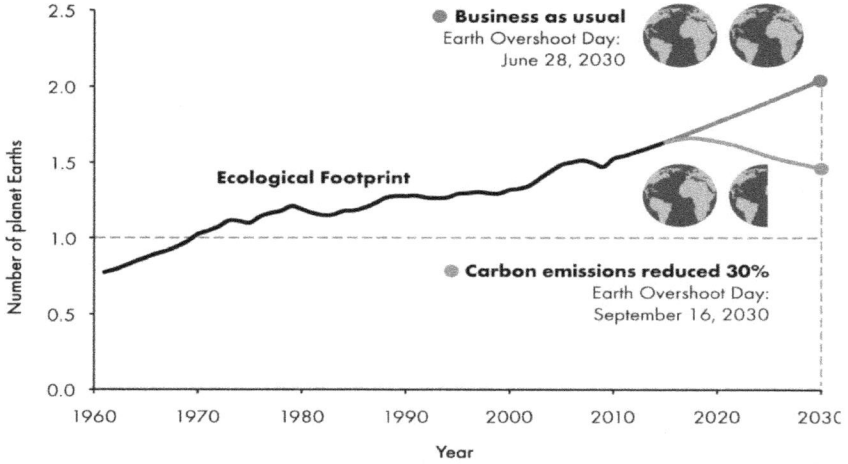

Figure 1.3: How many Earths does it take to support humanity?

Source: http://www.footprintnetwork.org/

In electrical engineering, overshoot is a common phenomenon. Basically, it means to go past a limit inadvertently. When you turn a knob too much in a meter, signals overshoot. When your car slides past stop sign on an icy road, your car overshoots. When you eat too much in a seafood buffet, and you feel sick the next day, you overshoot. Basically, we experience overshoot every day.

But, how about overshooting in ecology.

Let me give you an example. On August, 20, 1944, twenty-nine reindeers were brought to St. Matthew Island. Initially, there were abundant food sources, and the number increased exponentially. However, because there were no predators to cull the population, twenty years after the reindeers were introduced, they overshot the food carrying capacity of the island. And all of a sudden, massive population die off happened. The population of the reindeers plunged. In the end, ninety-nine percent of reindeers died of starvation.[3]

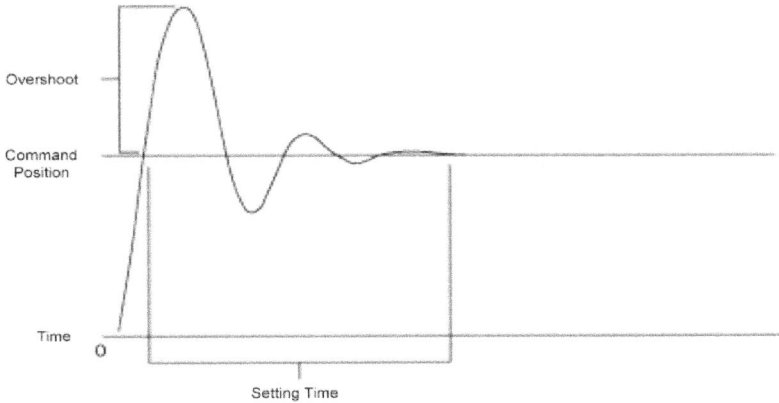

Figure 1.4: Overshoot

Source: Author

Overshooting is a natural phenomenon in all fields.

The aftermath of an overshoot is an inevitable collapse, followed by undershoot, before it adjusts back to seek its equilibrium. No one likes to see that happen to human population. Unfortunately, population growth is a trend likely to continue until a level out happens. The only way the world can reduce the severity of that collapse is add resistance (i.e., policies) to the current population growth.

Before tackling the problem of population growth in a moral way, first, we need to understand what has been fueling it.

What is fueling the world population growth?

If we look at the world population today, China, the country with the highest population in the world, has reached 1.44 billion in population[4], which accounts for 18.47% of the world's population. India, the second-largest population in the world, has reached 1.38 billion in population, which accounts for 17.7% of the world population.

The Most Populous Nations on Earth

Share of the world population by country (2018)

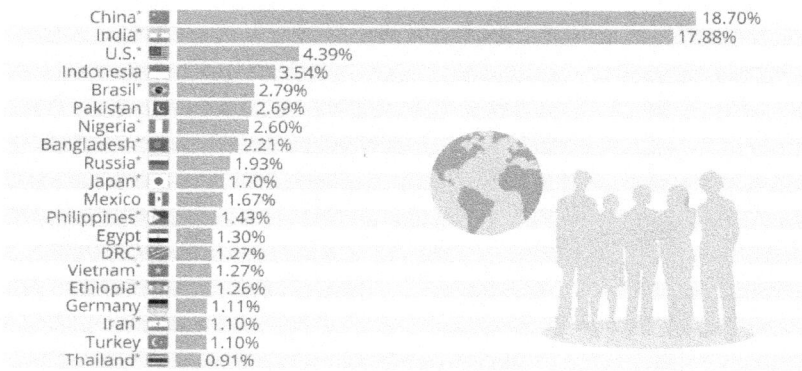

China*	18.70%
India*	17.88%
U.S.*	4.39%
Indonesia	3.54%
Brasil*	2.79%
Pakistan	2.69%
Nigeria*	2.60%
Bangladesh*	2.21%
Russia*	1.93%
Japan*	1.70%
Mexico	1.67%
Philippines*	1.43%
Egypt	1.30%
DRC*	1.27%
Vietnam*	1.27%
Ethiopia*	1.26%
Germany	1.11%
Iran*	1.10%
Turkey	1.10%
Thailand*	0.91%

* countries with estimated population size
@StatistaCharts   Source: IMF

statista

Figure 1.5: The Most Populous Nation on Earth

Source: IMF

As you can see, a slight increase in population growth in China (i.e., ~ 0.40%) will add another 5 million people on the planet, whereas, in a country like Bahrain, even with a population growth of 3.68% will only add 60,000 people on the planet. Statistically speaking, without taking moral issues and politics into account, by having fertility control in the top thirty most populous countries in the world, we can control the population growth of 76.7% of the world's population.

So, what is the current fertility rate in the world?

According to world data, from 1950 to 2015, on average, the number of children per woman had dropped from 5 to 2.5.[5]

Children per woman

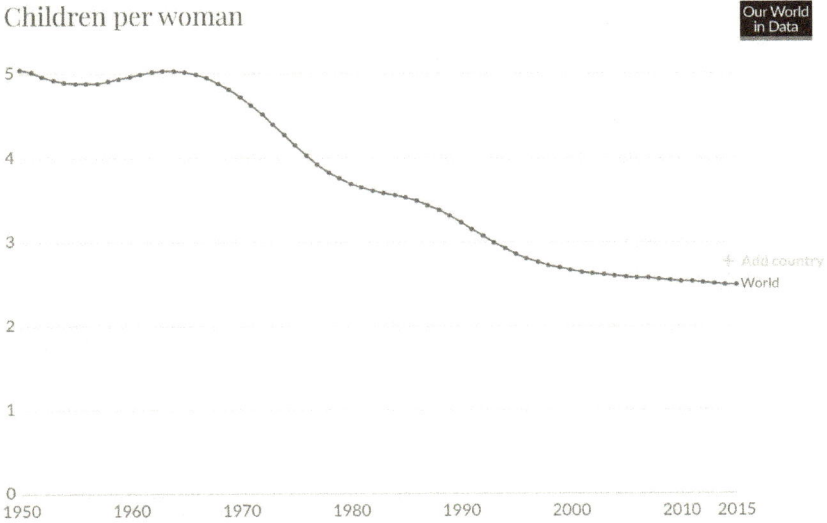

Source: UN Population Division (2017 Revision)
Note: Children per woman is measured as the total fertility rate, which is the number of children that would be born to the average woman
if she were to live to the end of her child-bearing years and give birth to children at the current age-specific fertility rates.

CC BY

Figure 1.6: Children per woman

Source: Our World in Data

Ironically, even with fertility rate dropping and family size shrinking, the world population growth of 1.05% (2020) is still making overshoot in population a predicament, rather than a problem.

If you look at the demographic of children born per woman in 2019, Africa, as a continent, has the highest birth rate. With a population of over 1.34 billion and a fertility rate of 4.4, this will be the source of population growth in the 21st century.

Children born per woman, 2019

Shown is the 'Total Fertility Rate' which measures the number of children that would be born to a woman if she were to live to the end of her childbearing years and bear children in accordance with the age-specific fertility rates of the specific year.

Our World
in Data

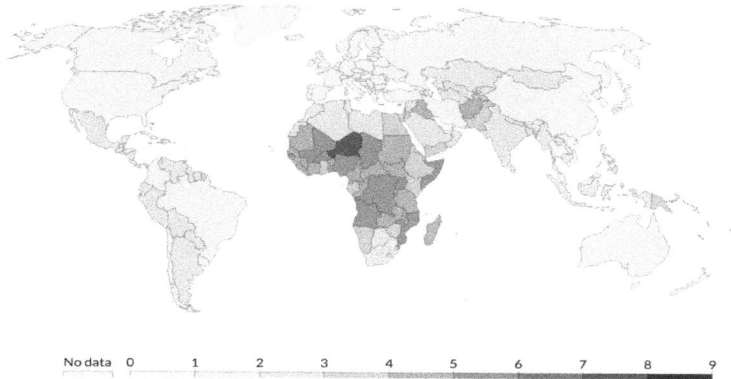

No data   0   1   2   3   4   5   6   7   8   9

Figure 1.7: Demographic of children born per woman 2019

Source: Our World in Data

Beside birth rates, human population is also related to the death rate. Due to the improvement in health care, people have longer life expectancy today. If you look at the global birth rates and death rates, it is interesting to see that both rates today have halved compared to the 50s.[6]

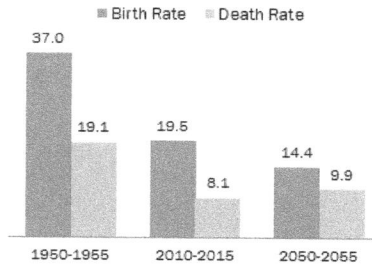

Global Birth Rates and Death Rates, 1950 to 2055

Per 1,000 people

■ Birth Rate   ▪ Death Rate

| | 37.0 | | 19.5 | | 14.4 |
| 1950-1955 | 19.1 | 2010-2015 | 8.1 | 2050-2055 | 9.9 |

Source: United Nations, Department of Economic and Social Affairs, *World Population Prospects: 2012 Revision*, June 2013, http://esa.un.org/unpd/wpp/index.htm

PEW RESEARCH CENTER

Figure 1.8: Global Birth Rates and Death Rates

Source: UN, Department of Economics and Social Affrairs

Population Growth Equation

A classic attempt to explain the human population and its impact on the environment is the IPAT equation. This equation was developed by biologist Paul Ehrlich and environmental scientist John Holdren in 1971. It tells us that a human's impact on the environment depends on population growth, affluence and technology.[7]

$I = P \times A \times T$

Where

I = ecosystem

P = Population

A = Affluence

T = Technology

It is common sense to know that when we have more people on Earth, we will have a higher demand for food, water and electricity. But population alone is not the sole factor that impacts the environment. For example, if person A lives in a village in a third world country where there is no electricity or automobiles, he is not going to leave much carbon footprint. On the other hand, if person B lives in a developed country like the U.S., where he drives to work every day, he is going to consume more energy. Affluence refers to the consumption by a country's population. That affluence will be magnified by the use of technology.

In order to reduce the impact on the environment, we must reduce our affluence by reducing our consumption. And consumption is something we can all control. It should be a priority for the world to allocate resources to develop technology that requires fewer resources, energy and waste discharge. Because we are all dependent on technology, its role in environmental impact is crucial for our survival.

By controlling 'P,' 'A,' and 'T' with policy, education and disciplines, the IPAT equation has given us a roadmap to design a sustainable future. It restrains us from overexploiting natural resources so that we can seek a balance and live within Mother Nature's limit.

Race to the Next Billion

Yet, in reality, at the current projection, the global population will reach 8 billion by 2023 – 2024.[8]

There isn't much time left to avoid an overshoot in human population.

We cannot afford to add a billion population every twelve years. With more and more people added to the planet demanding a better lifestyle, we will be creating a future with increasing strain on natural resources, shortage of freshwater, more inequality, and more natural disasters.

As I am writing now, Australia has just experienced the worse bushfires in history.[9] Venice was flooded by the highest tide in late 2019.[10] Overcrowding causes huge problems, such as housing, congestion, energy tension, air pollution and social problems. These are signs that we are hitting our sustainability limit because of the impact we did to our environment.

Is that the kind of future we want our next generation to live in?

Would we rather have business as usual and pretend nothing happened? Or would we rather act to take population growth a bit more seriously?

Chapter 2
The Story of Energy

"It is the economic, stupid" was a phrase used in Bill Clinton's 1992 presidential campaign against George. H. W. Bush. The message is simple - economic growth. Politicians always look at the world through the lens of economic growth as if it is the only thing they should look at, and other things that keep the world going will just take care of themselves.

This is far from the truth.

One of the main reasons we experienced remarkable economic growth in the last two centuries is our ability to exploit a cheap, abundant energy source called fossil fuel. It was like a treasure in our basement. It has transformed our society and personal lives in a profound way. Without it, everything happening in our society today would slow down dramatically.

In reality, economic growth has a limit. It is not defined by the monetary policy of the Federal Reserve or central banks around the world or what the interest rate is going to be so that business can borrow more.

It is also defined by energy.

At present, on the one hand, we have a fantasy economic system that is designed to grow forever. But, on the other hand, we have a realistic energy system and environmental constraint that can no longer meet that kind of economic growth. As we are heading into the mid- 21st century, we can feel this trend will only become more and more apparent.

11

This chapter is about the story of energy. It is probably one of the most important topics in this book. In my view, right now, the world should be laser-focused on bringing down global emissions by transiting away from fossil fuels. Yet, despite global energy policies and initiatives favoring renewable energy, there are a lot of important details missing in how this transition can be done realistically.

After all, can we fly a plane that carries 400 passengers using clean energy? Are battery-powered electric cars truly zero-emission? Can we maintain the same industry capacity of making steel and concrete using direct clean energy not derived from fossil fuel? How to make clean energy sustainable?

Before going into the details, we need to understand where we are today with the story of energy. But, for now, I would like to fly you back in time to see the earliest form of energy, and how it has shaped our society to the point we are today. Then, I will go into the current world energy outlook and forecast what the future of energy might look like in the 21st century.

A Brief Story of Energy

Imagine we've travelled back in time in a time capsule, a few hundred feet above the earliest human civilization. Our bird's-eye view would reveal layer after layer of dense forest with few clearings. Among those clearings, we would see clusters of huts and indigenous people encircling a fireplace with rising smoke.

In the earliest time, wood was humankind's very first source of energy. It was abundant. An ocean of forest was usually within sight in every town. It was a principal fuel that was generally used for heating and cooking.

While wood was the primary fuel back then, it wasn't the only fuel.[1]

As civilization starts to happen, humans began to learn how to discover alternative energy sources for lighting and transport as well. Besides wood, wax was another fuel that was used for lighting.

In about 300 BC, humans already know how to harness the energy of flowing water and convert it into useful forms of power. The first horizontal-wheeled mill was invented in Byzantium. The vertical-wheeled mill was invented afterwards to draw water from the river to irrigate farmlands.[2] The geared-wheeled mill further allowed the most generalized exploitation of waterpower.

Beside waterpower, wind energy has been used for thousands of years. People used wind energy to propel boats along the Nile River as early as 5,000 BC.[3] Originally, windmills and watermills were used for grinding grain. Over time, they were gradually refined for more and more applications.

In medieval Europe, wood was primarily the energy source for construction and heating.[4] But, due to deforestation, a wood shortage began to emerge by the second half of the 16th century. Countries, such as Britain, began to resort to coal for heating. Even so, the existence of coal as an energy source was no secret back then. Coal was one of our earliest sources of heat and light. In the East, the Chinese were known to have used coal more than 3,000 years ago.[5] Even so, coal was known as an inferior fuel because it is dirty. It wasn't widely used because the extraction of coal was dangerous and environmentally unfriendly. Also, the mining technology was not mature enough for coal to be mined and transported out of the mines for use. That was why coal was largely ignored.

By the 17th century, coal had revolutionized beyond home heating. Thomas Newcomen, an English inventor, had created the first atmospheric engine in 1712 – a useful steam engine to pump water from the bottom of the mines by burning coal.[6] Never again did men need to harness power from labor, livestock and water, which are limited and restrictive. Later on, the famous Scottish inventor, James Watt, discovered a way to improve Newcomen's design by increasing efficiency with less coal.[7] By the early 1800s, Watt's steam engines were used in factories throughout England, which helped to power the first Industrial Revolution. It is worth noting that before steam power, most factories and mills were powered by wind, water, labor or livestock. The drawback of renewable muscle power is that it is very limited. Waterpower was good, but it limited the geographical location of factories. On the other hand, steam power was like large portable batteries that allow factories to be built anywhere. It was reliable to power very large machines. Coal, steam engines and machines, all born in the Industrial Revolution, revolutionized our economy and society as a whole. First, it was England, then America, Germany and the rest of the world. This led to more and more people working in factories. In fact, the origins of the eight-hour workday we have today can be traced back to the Industrial Revolution.[8] An economy previously based on local consumption had been transformed entirely by coal. This is the world's first energy revolution.

[Note: A steam engine uses hot steam from boiling water to drive a piston back and forth. The movement of the piston was then used to power a machine or turn a wheel. To create steam, most steam engines heated water by burning coal.]

While the first industrial revolution (1760 to 1840) gave rise to coal, iron, railroad and the textile industry, it was not until the beginning of the

19th century before humans began the second energy revolution. When Colonel Edwin Drake drilled the first successful oil well in Titusville, Pennsylvania, in 1859, no one had any idea how petroleum would change the world in the future.[9] Like coal, petroleum was only used as a lubricant and fuel for lamps. Because early industries used whale oil for oil lamps, whale had been hunted close to extinction.[10] However, the oil market did not have a smooth ride. Oil had lost its primary market when Thomas Edison invented the light bulb and created the electrical generation industry. The demand for oil peaked and began to decline.[11]

Meanwhile, an American business magnate called John D. Rockefeller began his purchase of crude oil in Pennsylvania, Ohio and West Virginia, and refined it under the business Standard Oil.[12] It developed its own distribution and production system in addition to refineries. By 1880, Standard Oil became a worldwide monopoly of petroleum. It controlled 90% of the U.S. oil business, as well as oil business in the rest of the world.[13]

Despite Edison's electricity breakthrough that caused the primary market for oil to disappear, the invention of the automobile reversed that trend. Besides that, in the chemical industry, new synthetic material using petroleum-like plastic and nylon had also caused the demand of oil to surge. This completely transformed the world economy.[14] Since then, everything in the 20th century was attributed to the use of petroleum. A profound revolution had been seen in agriculture, aeronautics and warfare. WWI, in particular, changed the role of petroleum in the world.[15] As a result of oil, motorized transport, tanks and airplanes changed the nature of war dramatically. Oil had clearly given an unfair advantage to countries that had access to it. It became a strategic material.

Apart from oil, another type of fossil fuel is natural gas. Ironically, when natural gas was first discovered, it was a byproduct of coalmine and oil production. Most oil wells discovered also had natural gas nearby. However, natural gas was considered to have no value and was regarded as a dangerous waste product. When natural gas was first used commercially, it was used exclusively as fuel for street lamps.[16] No pipeline existed to transport gas over long distances. But with the rise of the electricity market, many gas producers were forced to look at other markets. The development of the pipeline network in the 50s and 60s constructed throughout the U.S. allowed easy distribution of natural gas, and, therefore, it was economically attractive.[17] Since then, natural gas has become a versatile form of energy we use in our daily lives.

The Journey to Electrifying the World

Besides oil, one of the most profound discoveries that reshaped the 20th century is electricity. Unlike fossil fuels, electricity is not an energy source. It has no physical form. If you remember your science lesson, an atom constitutes subatomic particles called electrons, protons and neutrons. Electricity is the flow of electrons in atoms. We call it electric current. A unit of electric current is an Ampere (SI unit is A), meaning one coulomb of charge per second. Electric current is the carrier of energy that makes it easy to access and use. We call this electricity.

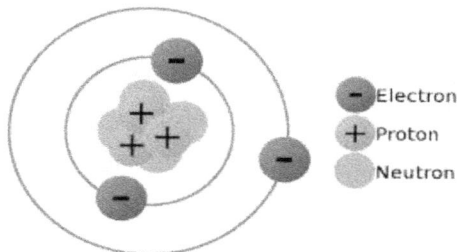

Figure 2.1: Inside an atom

Source: Author

But who discovered electricity?

Although electricity has been around for an eternity, we have only harnessed its power for 250 years.

Flying a kite in a thunderstorm was Benjamin Franklin's (1706-1790) most famous experiment.[18] It led to the invention of the lightning rod and the understanding of positive and negative charges. Benjamin's kite experiment demonstrated that lightning was an electrical discharge. By flying a kite with a key tied to the end with string connected to a Leyden jar, the electric current traveled down through the kite into the Leyden jar when lightning struck it.[19]

Figure 2.2: Benjamin's kite experiment

Source: Benjamin Franklin Historical Society

[Note: A Leyden jar is an old battery consisting of a glass jar wrapped in lead foil with lining going into the jar as a metal rod. Electricity could be stored for a short period in a Leyden jar.]

Perhaps one of the first major breakthroughs in electricity happened in 1831. A British scientist, Michael Faraday (1791 – 1867), discovered the principles of electricity generation.[20] He observed that by moving magnets inside the helix of copper wires, he could induce an electric current. In other words, if we have motions in a magnetic field, we can generate electricity. This major discovery of electromagnetic induction changed the way we use energy. It also produced the basis of rotating electric motors.

Faraday's achievement in 1831 gradually led to the production of practical machines that converted mechanical power into electrical energy. A dynamo was the first Direct current (DC) electrical generator invented to deliver power for industry. It has a stator that provides a constant magnetic field, and a set of rotating windings called armature that turn within that field.[21] In Faraday's law, the motion of wire in the magnetic field creates an electromotive force that pushes the electrons on the windings and creates current. Moving the magnet back and forth produces alternating current (AC). But, back then, everyone believed that AC was useless. Therefore, in the dynamo, a component called the commuter was used to force the AC back to DC, and dynamos driven by steam engines were used in power stations.

Virtually, all electric power is produced using Faraday's principle. Whether it is the wind, hydro, gas, coal or oil, all the fuels are used to create motions through a generator to generate electrical current.

But how about electricity transmission?

Between 1870 and 1880, the DC power system dominated. Thomas Edison devised the first public electric power transmission system in 1880 using direct current. But the DC power system only powered factories and cities where distances were small. It could not reach 95%

of the population.[22] To send DC power over long distances needed high voltage because the electrical resistance of the transmission lines between pylons increases over distance. In reality, not all of the power delivered could reach the household. A fair amount of it will be loss due to heating on cables.

In late 1880, Thomas Edison and Nikola Tesla were embroiled in a battle known as the *War of Currents*.[23] Thomas Edison's DC – a current that continuously runs in one direction was the U.S. standard; however, DC faces the problem of power transmission. Tesla, on the other hand, thought that the way to overcome this was by using AC. To avoid power loss, in an AC power transmission system, voltage is first stepped up by the use of transformers to minimize the current and reduced power lost along the transmission line, then the voltage is stepped back down again when it reached the end users.

[Note: DC is direct current, current that flows in one direction only. AC is alternating current, electric current that periodically reverses direction.]

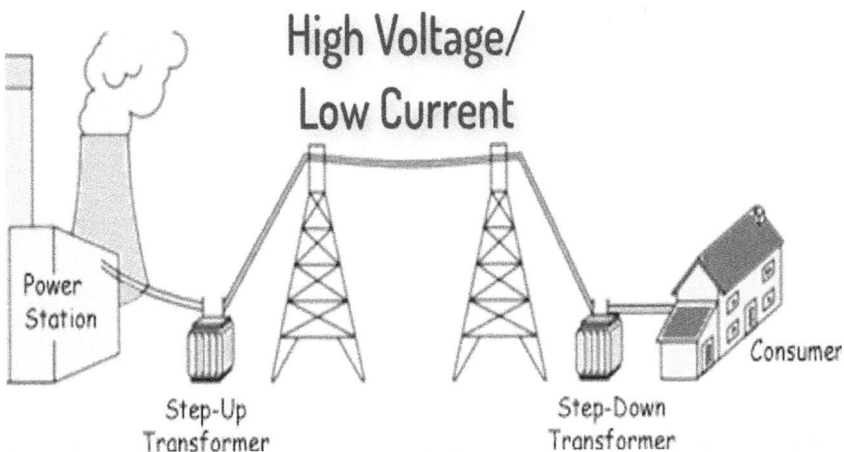

Figure 2.3: Power Distribution

Source: Author

But Thomas Edison doesn't want to lose the royalty he earned from DC patents. So he discredits AC and misinforms the public about how dangerous AC is.[24] He tried to prove this by publicly killing animals with both DC and AC.[25] But, in the end, the current war did not turn in his favor.

In 1886, George Westinghouse Jr., aided by Nikola Tesla, successfully developed the AC power transmission model that can operate more economically over long distances. In 1893, Westinghouse won the bid to supply electrical power for the World's Columbian Exposition and won the major part of the contract to build the Niagara Falls hydroelectric project later that year.[26] That event marked the standardization of AC in grid power in the world today.

Even though AC has dominated the world energy market in the 19th and 20th centuries, the current battle is far from over. Later on in this chapter, I will explain how high voltage direct current (HVDC) is about to make a comeback in the mid-late 21st century.

***

So, you see, if it weren't for fossil fuel and electricity, we would probably still be living like our 18th-century ancestors. But fossil fuel is non-renewable and is scarce. To understand the story of energy in the 21st century, the main question we should be asking is, what is the world's energy outlook today?

World Energy Outlook

Today, everyone talks about going renewable. Everyone talks about saving our environment by reducing our carbon footprint by replacing fossil fuel. Companies have been boosting growth in renewable energy efficiency. Global investment in new renewable energy capacity is set to reach USD 2.6 trillion by the end of 2019.[27] Despite the noble intention and the great effort, the reality shows a very different picture.

Below is a chart showing the global primary energy consumption in the world in 2019. It is the big picture, showing what type of energy source the world is consuming as a whole.

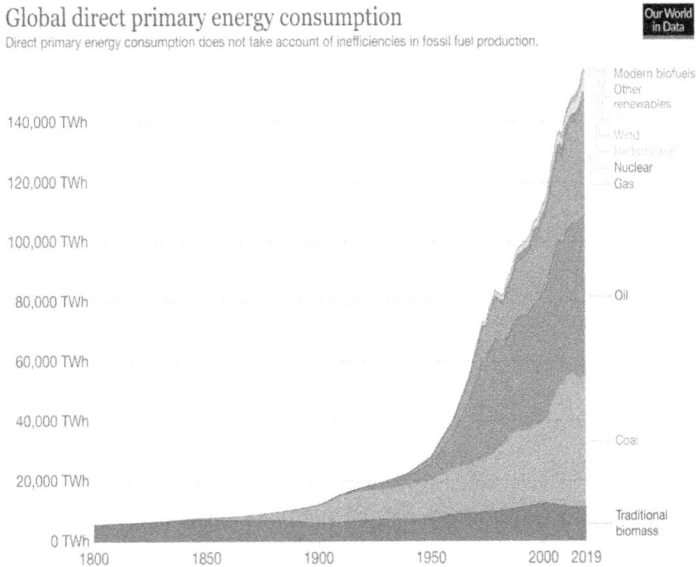

Figure 2.4 Global Primary Energy Consumption
Source: Our World Data

If you look at the global primary energy consumption, it may surprise you that despite all the hype about going solar, solar energy consumption only accounts for 0.37% of total energy consumption in the world. For wind, it only account for 0.81%. On the other hand, more than 80% of the world's energy consumption comes from fossil fuel.

2018		2018	
Other renewables	625.81 TWh	Other renewables	0.4%
Solar	584.63 TWh	Solar	0.37%
Wind	1,269.95 TWh	Wind	0.81%
Nuclear	7,109.03 TWh	Nuclear	1.72%
Hydropower	4,193.10 TWh	Hydropower	2.67%
Natural gas	38,488.57 TWh	Natural gas	24.51%
Crude oil	54,219.68 TWh	Crude oil	34.52%
Coal	43,869.48 TWh	Coal	27.93%
Traditional biofuels	11,111.11 TWh	Traditional biofuels	7.07%
Total	161,471.36 TWh	Total	100%

Figure 2.5 Global Energy Consumption by energy type

Source: Our World Data

[Note: 1TWh is 1,000,000,000kWh]

Do you begin to see how the numbers don't add up despite the world's increasing investment in renewable energy? According to UN figures in 2015, global investment in solar power, wind turbines and other renewable forms of energy was $266 billion [28], but renewable energy sources deliver just 10.3% of global electrical power.

While people are excited about the recent growth rates in renewable energy, they fail to realize that even a tiny percentage growth in fossil fuel will dwarf a large percentage gain in renewables.

The reality is our economy is a function of oil. Despite the media hype, the world will still heavily depend on crude oil in the 21st century.

But what is the story of oil?

Who Controls the Oil Price?

In the first few decades of the 20th century, the U.S. was in a position to control the price of oil in the world. This changed in the second half of the century when U.S. production declined while the Middle East's production increased.

Back in the 20th century, the U.S. was simultaneously the world's largest consumer and oil producer. Its economy grew tremendously from using petroleum in mass production of automobiles and developing the commercial airline industry. As a result of Standard Oil's monopoly and ruthless and illegal business methods, in 1911, a lawsuit was bought against it by the U.S. government in 1906 under the Sherman Antitrust Act of 1890 to break it down into 34 separate companies.[29] The intention of this lawsuit was for these companies to compete with one another. In reality, the companies had little real incentive to compete. They acted together in setting prices for a decade or more. From approximately World War I to 1970, the three largest post-breakup companies, Standard Oil of New Jersey (Exxon), Standard Oil of New York (Mobil), and Standard Oil of California (Chevron), joined with Gulf, Texaco, BP, and Shell to form a cartel, earning them the nickname the "Seven Sisters."[30] These seven companies owned the vast majority of the world's oil and controlled the economic fate of entire nations.

Preceding the 1973 oil crisis, the Seven Sisters controlled 85% of the world's oil reserves. But, as more Middle East reserves were discovered, countries around that area slowly took over, brought down the Seven Sisters, and became the largest oil producer.[31]

The Organization of the Petroleum Exporting Countries (OPEC) is an intergovernmental organization created at the Baghdad conference on 10-14 September 1960 by Iran, Iraq, Kuwait, Saudi Arabia and Venezuela.

Currently, it has fourteen member countries, including Algeria, Angola, Congo, Ecuador, Equatorial Guinea, Gabon, Iran, Iraq, Kuwait, Libya, Nigeria, Saudi Arabia, United Arab Emirates, Venezuela. OPEC provided 40 percent of the world's crude oil. Its export represents 60 percent of the total petroleum traded in the international market. Because of its great market share, OPEC can influence international oil prices. As the largest oil producer within OPEC and the world's largest oil exporter, Saudi Arabia can influence oil prices.[32]

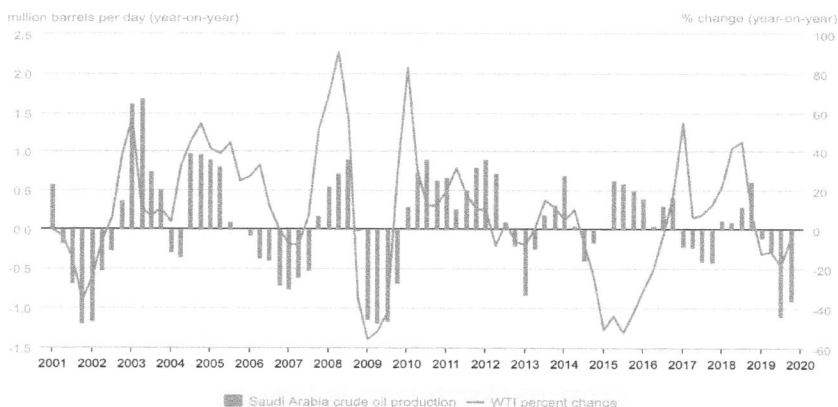

Changes in Saudi Arabia crude oil production can affect oil prices

Changes in Saudi Arabia crude oil production and WTI crude oil prices

Source: U.S. Energy Information Administration, Refinitiv

Figure 2.6 Change in Saudi Arabia crude oil production can affect oil price

Source: EIA

What Determines the Spot Price of Oil?

The worldwide spot price of oil depends on the supply of oil from Non-OPEC and OPEC, the demand of OECD and non-OECD, the balance (i.e. the inventories stored for future use), and the financial markets. Geopolitical events like 9-11, Arab oil embargos in 1973 to 1973, the Iran-

Iraq war in the late 90s and the Asia financial crisis in 1997 can also lead to actual disruption or future uncertainty of the demand and supply of oil.

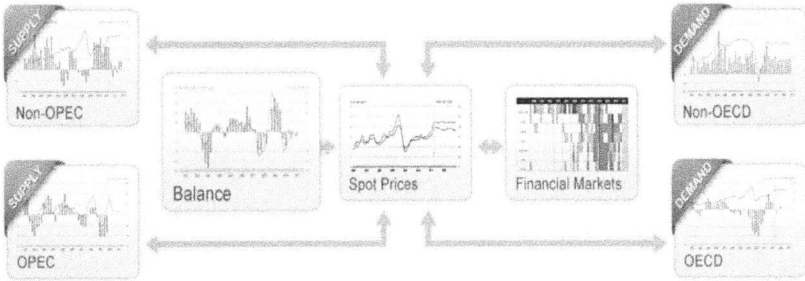

Figure 2.7 Changes in Saudi Arabia crude oil production

Source: U.S. Energy Information Administration

What is the Story of Oil?

In 2019, the world total oil production is 100.63 million barrels per day.[33] The top three oil producers are the U.S., Saudi Arabia and Russia.

The 10 largest oil[1] producers and share of total world oil production[2] in 2019[3]

Country	Million barrels per day	Share of world total
United States	19.51	19%
Saudi Arabia	11.81	12%
Russia	11.49	11%
Canada	5.50	5%
China	4.89	5%
Iraq	4.74	5%
United Arab Emirates	4.01	4%
Brazil	3.67	4%
Iran	3.19	3%
Kuwait	2.94	3%
Total top 10	71.76	71%
World total	100.63	

[1] *Oil* includes crude oil, all other petroleum liquids, and biofuels.
[2] Production includes domestic production of crude oil, all other petroleum liquids, biofuels, and refinery processing gain.
[3] Most recent year for which data are available when this FAQ was updated.

Figure 2.8: The largest oil producers and share of total world production in 2019

Source: U.S. Energy Information Administration

On the consumption side, the world total oil consumption was 98.76 million barrels a day.[34] U.S., China and India made up more than a third of the world's oil consumption.

The 10 largest oil[1] consumers and share of total world oil consumption in 2017[2]

Country	Million barrels per day	Share of world total
United States	19.96	20%
China	13.57	14%
India	4.34	4%
Japan	3.92	4%
Russia	3.69	4%
Saudi Arabia	3.33	3%
Brazil	3.03	3%
South Korea	2.63	3%
Germany	2.45	2%
Canada	2.42	2%
Total top 10	59.33	60%
World total	98.76	

[1] *Oil* includes crude oil, all other petroleum liquids, and biofuels.
[2] Most recent year for which data are available when this FAQ was updated.

Figure 2.9: The largest oil consumer and share of total world consumption in 2017

Source: U.S. Energy Information Administration

As I am writing now, the world has 1.65 trillion barrels of proven reserve.[35] At the current rate of consumption, it is estimated that the world will have 47 years of oil left.[36]

History of World's Proven Oil Reserves

**Proven Oil Reserves
(barrels)**

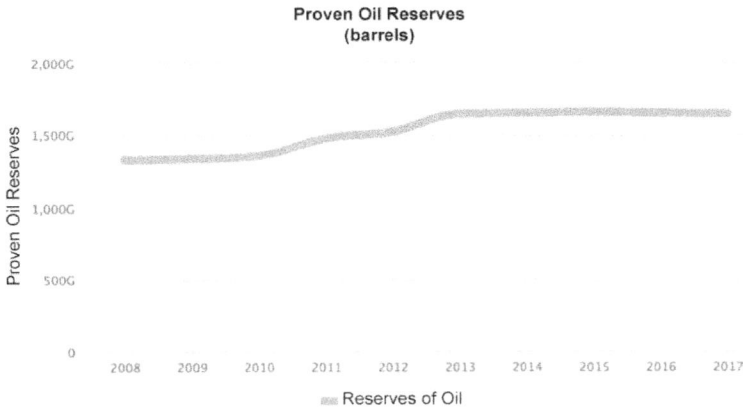

Figure 2.10: Proven Oil Reserve in 2017

Source: U.S. Energy Information Administration

[Note: Venezuela has the largest amount of oil reserves in the world with 300.9 billion barrels, 18.2% of the world's reserve.[37] Saudi Arabia has the second-largest amount of oil reserves in the world with 266.5 billion barrels, 16.2% of the world's reserve.[38]]

Peak Oil

In 1956, geologist M. King Hubbert presented a formal theory called Peak Oil. Hubbert's peak theory says that for any given geographical area, from an individual oil-producing region to the planet as a whole, the rate of petroleum production tends to follow a bell-shaped curve.[39]

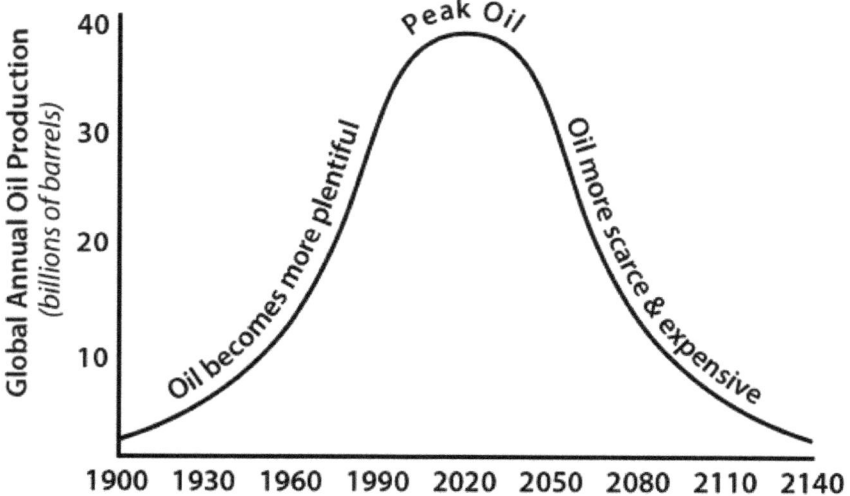

Figure 2.11: Hubbert Curve of Global Oil Production

Source: www.explainingthefuture.com

Peak oil is not synonymous with running out of oil. It means oil is cheap to extract on the way up, and the reverse is true on the way down. Each barrel will be more costly in terms of time, money and energy to extract. Eventually, it costs more to extract a barrel of oil than it is worth, which is the point when an oil field is abandoned.

Take U.S. Field production of crude oil as an example, from the first oil field drilled in 1859 in Titus, Pennsylvania until 1970, more and more oil was progressively pumped out of the ground.[40] But beyond 1970, the trend has been declining.[41]

U.S. Field Production of Crude Oil

Thousand Barrels per Day

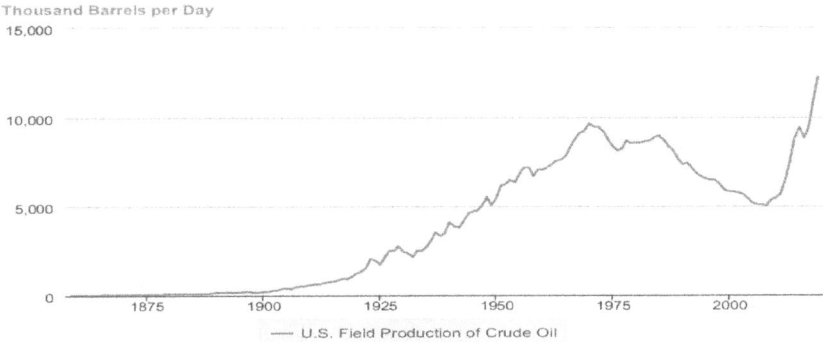

Figure 2.12: U.S. Production of Crude Oil

Source: EIA

[Note: If you look at the graph carefully, you will see a sudden sharp rise in crude oil production in the recent decade. So, does that contradict Hubbert's Peak Oil theory? The rise of oil production in recent decades is due to fracking technology and horizontal drilling. However, fracking cannot return us to the cheap energy era. We will cover that later.]

Can we discover more oil fields?

In the oil industry, there is a saying – in order to pump oil, you have to find it first. The grim reality is that when we look at the worldwide oil discovery trend, you will discover that there has been a steady increase in oil field discovery in every decade up to 1964. Beyond that, oil discovery has been declining steadily every decade.[42]

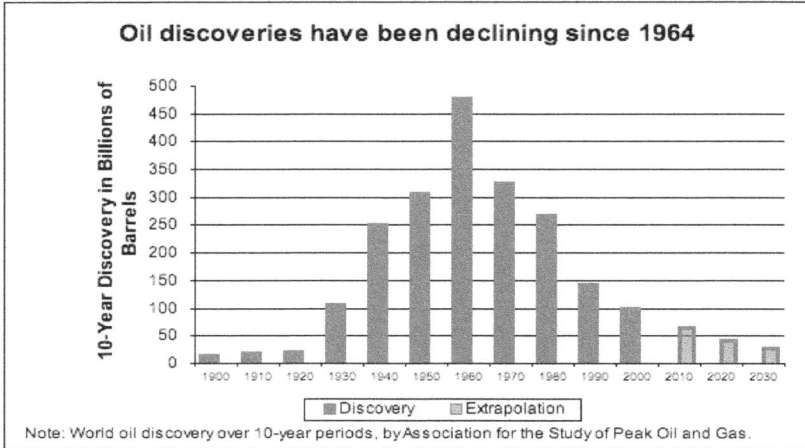

Figure 2.13: U.S. Oil Discovery have been declining since 1964
Source: www.uranium-stocks.net

Energy Invest Energy Return

One of the fundamental concepts energy policymakers least talk about is it takes energy to produce energy. If you think in terms of energy but not dollars, everything becomes crystal clear. To produce a wind turbine, it takes energy to manufacture the wind turbine blades, the shaft, the gearbox and the generator. Then it takes energy to transport these parts to a wind farm. It takes energy for people to come to work. It takes energy to power bulldozers and cranes for the installation.

To put this into perspective, below is a figure showing the net energy cliff. It shows the energy used in production (red) to gain energy available for consumption (green). We call it energy return on energy invested (EROEI).[43]

THE NET ENERGY CLIFF

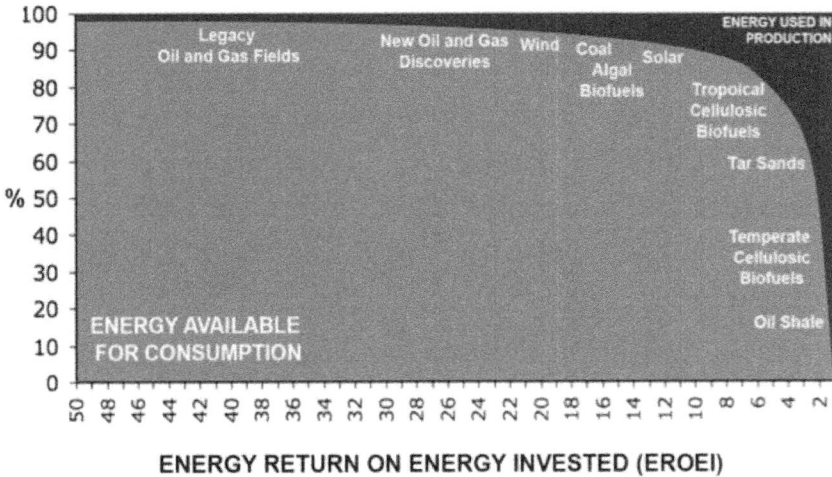

Figure 2.14: Energy Cliff

Source: wikimedia.org

In a nutshell, EROEI is an energy budget. In 1930, for every barrel of oil we used to find oil, we produced 100 barrel of oil in return. It has a very high EROEI. However, by 1970, the return gets smaller because fields were a lot smaller, and it is deeper to extract available oil. Today, it is about 3:1.

So, when the EROEI of an energy source is less than or equal to 1, it means that the energy source becomes an energy sinker. It takes more energy to get that unit of energy source. A viable energy source needs to give an EROEI of at least 3:1. If you look back at Figure 2.9, the main reason why we saw a sudden uptrend of oil production in the U.S. is largely due to shale oil using unconventional production methods like fracking.[44] It only has an EROEI of 1.5:1. So, despite a soaring oil production rate from fracking technology, the EROEI is far too small. It produces too much greenhouse gas, pollution and environmental issues.

If we are to sail through the 21st century, we will have to master the concept of net energy and invest our energy to transit to renewable energy more wisely. We cannot be blindfolded by the price of energy, as it is irrelevant. Money is an illusion. EROEI is our reality. The real driving force behind our economy is under the assumption that we have more and more available energy each year. The bottom line is that it takes non-renewable energy to create renewable energy. That's why we need a reality check on our energy transition roadmap to renewable energy in the 21st century.

[Note: In reality, although the EROEI formula is quite simple, the factors for calculating the EROEI differ greatly in the industry and research papers. There are a lot of subjectivities. Take the EROEI of photovoltaic (PV) as an example; some papers propose to calculate the energy input side of PV without taking into account the energy invested into necessary labor, the energy embodied into the materials to make a PV and associated accessories like batteries etc. To me, this is erroneous and misleading. This kind of assumption will lead to bad energy policy and turn something that is supposed to be energy sink into something that gives energy gain.]

A Reality Check on our Transition to Renewable Energy

Imagine you are building a hydroelectric dam to generate electricity through hydropower? You need to build turbines and dams. The majority of these constructions intake at least some form of non-renewable energy. (e.g., workers driving to work by car need petrol; construction requires bulldozers that need petrol). The energy intake through labor and capital should be taken into account in the energy equation.

If the energy payback time of that one unit of renewable energy is far too long (e.g., 10+ years), then the entire concept of renewable energy is actually depleting our non-renewable energy at a much faster pace.

I am not against using renewable energy, but we need to invest in the type of renewable energy that offers high-energy return or improve the energy return on how we generate existing renewable energy.

Below are a few reality checks on our transition to renewable energy. I will talk about the potential problems with each existing type of renewable energy, the possible ways to improve on the energy return, as well as some novel types of renewable energy. Let's start with solar energy.

Solar Energy

The source of solar energy comes directly from the sun. To gather that energy, convert it into a useful form, and store it for future use when the sun is not shining is the challenge solar energy faces. A large energy storage system is needed to provide a constant, reliable source of electricity when a cloud goes overhead.

So, a shift to solar energy translates to a huge demand in an energy storage system, such as a battery. Among other battery types, a conventional battery like the lithium-ion battery is the best option for solar panels because of its efficiency.[45] But the main problem is that batteries, in general, have a shelf life. It will need replacing after a certain number of years. And lithium-ion batteries are not very recyclable. The chemicals inside lithium-ion batteries like cobalt, nickel, and manganese, are considered toxic heavy metal.[46]

Solar energy might be clean energy of its own. But the solar panels on our rooftop only last for twenty years, and the battery lasts about a decade. So, if we look at the whole picture, you will realize that once the solar panel and batteries reach the end of their life, there will be a mountain of hazardous waste. It is a looming waste crisis waiting to explode.

So, are there other ways to store energy besides conventional battery?

Hydrogen

Hydrogen is the most abundant element on the planet. Unlike the lithium-ion battery, hydrogen does not produce any pollutants. Hydrogen is clean, renewable and efficient. A hydrogen-powered fuel cell solves just about any environmental problems that a conventional battery brings.

So, when sunlight is converted into electricity in a solar cell, instead of using a battery as a medium to store this energy, it uses hydrogen. The photovoltaic electricity will undergo electrolysis of water to produce hydrogen and oxygen. The hydrogen becomes a medium to store this energy. In order to convert it back into electricity again, this hydrogen will need to react with oxygen in a fuel cell.[47]

Currently, the conventional way to produce hydrogen in the industry is by burning natural gas like methane, but this process will produce hydrogen and carbon dioxide as well. While this conventional method might be undesirable, the byproduct of burning fossil fuel, hydrogen, is seen as a way for us to transit to a hydrogen economy, while maintaining the economy.

Although electrolysis has been around for many decades and is widely used to produce oxygen and hydrogen in the chemical industry, in terms of energy storage, hydrogen is still in the infancy stage of development. Hydrogen energy infrastructure is a lot different from our present energy infrastructure.[48] So, in order for the world to effectively transit from fossil fuel to renewable energy in a sustainable way, the world should heavily invest in technologies like hydrogen fuel cells, improve its conversion efficiency, and diversify into building more hydrogen energy infrastructure, while we still can.

Hydro power

According to IEA, hydropower represents about 17% of electricity production in the world.[49] Among other types of renewable energy, hydropower constitutes the largest renewable energy source.

Hydropower uses the water stored in dams and rivers to create electricity in hydropower plants. The falling of water rotates the blades of turbines, which spins a generator and converts the mechanical energy of the spinning turbine into electrical energy. Producing electricity using hydropower has a lot of key advantages. It is reliable, renewable, and the water that runs the power plant is from nature. Also, hydroelectric power is relatively easy to store during times of low demand. It is simply stored in the form of water.

However, hydroelectric power requires a very high set up cost, which is not favorable for developing countries. Also, hydroelectric power is not the perfect clean energy we once thought. Hydroelectric dams can produce significant amounts of carbon dioxide and methane, depending on the plant.[50] This is because a large amount of carbon tied up in trees and plants is released when the reservoir is initially flooded, and the plants rot. The decay of plant matter settles at the reservoir's bottom decomposes without oxygen, resulting in a build-up of dissolved methane. This is then released back to the atmosphere when water passes through the dam's turbines. At present, these greenhouse gases from a reservoir are not taken into account for greenhouse gas inventory.

Despite all these factors, hydropower has a very important role to play in our energy transition to renewable in the 21st century. It has the highest EROEI among all renewable energy.[51] In fact, in 2018, hydropower in China had grown 20-fold to a capacity of 352GW, representing one-quarter of the world's installed capacity.[52] Since 2012, the Three Gorges

Dam in China became the world's largest hydroelectric plant.[53] It has an annual generation capacity of 101.6 TWh (2018).[54]

Wind Energy

Among all other renewable energy, wind energy is one of the fastest to develop globally. While wind energy can vary depending on the weather and time of day, it cannot be depleted. It is the cheapest source of large-scale renewable energy. Wind energy also has very good EROEI. Unlike hydroelectric power, where most of the regions suitable for hydro are thoroughly exploited, wind energy has room for expansion.

The key challenge for wind energy is that peak available for wind does not translate to peak energy demand. Hydrogen fuel cell technology, like electrolyzers, might be a good candidate to store this renewable energy.

However, wind energy is not without its drawbacks. Firstly, wind energy is not suitable for every country because it is limited by terrain and geographical location. Secondly, it cannot be standalone energy because of its uncertainty and must be integrated into existing power plants like coal to make up for the difference in situations where there's no wind. Moreover, wind power needs the construction of a new grid system to transport the electricity from where the wind farms operate.

As you might already realize, despite a relatively high EROEI of wind energy in theory, in reality, the solution of constructing a very large scale new grid system to support a wind farm will only translate into depleting our dwindling non-renewable energy much more quickly. The lifespan of typical wind turbines is 20 to 25 years.[55] Once the lifespan is reached, new wind turbines will be needed.

Ocean Energy

Ocean energy is probably the least talked about renewable energy. It refers to all forms of renewable energy derived from the sea: wave, tidal and ocean thermal.[56]

Wave and tidal energy are generated by converting the ocean waves and tidal into electricity. Tides are a renewable energy source that could provide power 18 to 22 hours a day. Tidal turbines are placed beneath the water, and the blades rotate according to the tide. The turbines then turn the tidal generator module to produce electricity. Underwater cables return the electricity for use ashore.

Ocean thermal energy conversion (OTEC) is a technology that uses ocean thermal gradient between the cooler and shallow parts of the ocean (800-1000m) to run a heat engine to produce mechanical work, such as turning a generator to produce electricity.[57] Generally, it operates with a temperature gradient, usually around 20°C, but the higher the temperature gradient, the more electricity it can create. It is clean. Unlike solar or wind, OTEC is reliable and can run 24/7 a day 365 days a year. Any nation surrounded by ocean can deploy this technology. It is by far my favorite form of renewable energy because of multi-use. An OTEC plant can also integrate with other desalination technology to produce freshwater. This can potentially solve the problems of countries suffering from clean water and a shortage of electricity.

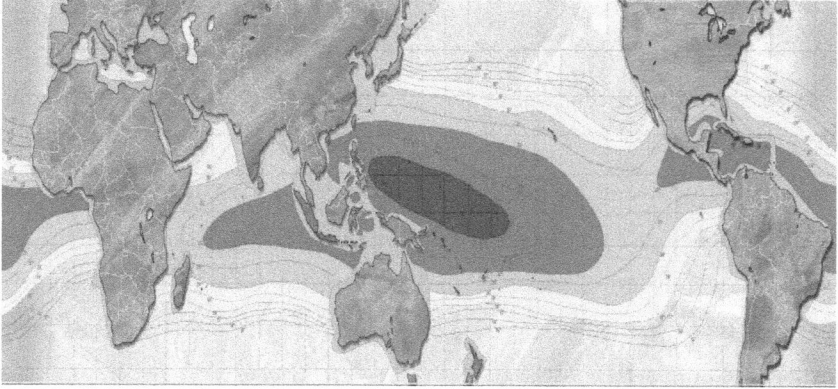

Figure 2.15: Distribution of OTEC Thermal Resources in the world

Source: U.S. Department of Energy

Although the efficiency of OTEC is very low (7%), this does not affect the feasibility of this technology because the fuel (i.e., water) is free.[58] However, OTEC's barrier is very high capital cost and not many countries have the expertise with experience in building this new technology. Right now, an OTEC plant built is still between 1MW to 10MW. So, say, an average Australian house consumes 6750 kW annually, a 1MW OTEC is just enough to power 148 Australian homes per year. So, the challenge of OTEC will be to scale it up massively. From an energy perspective, this type of technology will be vital if we are to migrate to a sustainable, renewable energy future.

Energy Efficiency

Apart from finding alternative renewable energy sources, one of the quickest and actionable things we can do today is make our electrical equipment smarter and more energy efficient.

Electric motors account for 45% of energy consumption in the world, and two-thirds of the industrial power consumption.[59] Lighting is only a distant number two, consuming about 19%. Motors run 24/7 a day, 365 days a year. Ironically, we only hear about improving the efficiency of lighting most of the time.

The truth is that the International Electrotechnical Commission (IEC) does require electric motor manufacturers to comply with energy efficiency. IEC 60034-30 classifies motors into different international efficiency (IE1 to IE4).[60] In Australia, Minimum Energy Performance Standards make it compulsory for motor manufacturers to comply with the energy requirement. An improvement in motor efficiency can save a significant amount of money each year, provided that the cost of electricity stays constant. Other technology like VSD can control the motor's speed, thus reducing the electricity consumption of the motor and enhancing efficiency.[61]

Global Energy Interconnection

The key challenge of the energy transition in the 21st century isn't just about energy generation alone. It is also about energy storage, energy transmission, energy distribution, global energy policy, global energy governance, environmental impacts and the overall transformation of the energy market. How do we integrate renewable energy into our grid? Could we make our grid system smarter? Is it possible for anyone in any country to sell renewable energy to one another P2P using block chain technology?

Global Energy Interconnection (GEI) is a novel concept to interconnect power systems around the globe.[62] It integrates large-scale renewable energy deployment using ultra-high voltage (UHV) grid to interconnect across countries, supported by a Smart Grid. UHV is the pivotal technology that supports long distance electrical transmission.[63] There are two options for UHV: Ultra High Voltage Direct Current (UHVDC) and Ultra High Voltage Alternating Current (UHVAC). UHVDC has smaller losses compared to UHVAC over long distances because it has no skin effect. The absence of inductance in DC means UHVDC has better voltage regulation. However, the starting capital is higher for UHVDC because it is more expensive for the terminal converter station, which is not required for UHVAC transmission. The break-even distance for UHVDC is around 600km. So, if the transmission distance is greater than 600km, UHVDC is more economical.[64] At the beginning of 2018, the State Grid Corporation of China (SGCC), the world's largest power company, began work on a 1.1 million volt transmission line. This will be the largest transmission line in the world, capable of delivering huge amounts of power over thousands of miles from Xinjiang to Anhui. A similar project, Yunnan–Guangdong HVDC, had been constructed from 2007 to 2010 using UHVDC technology to transmit electricity from hydroelectric plants from the Yunnan to Guangdong (1418 km). Yunnan–Guangdong HVDC was the first UHVDC link in the world operating at transmission of 800kV.[65] This is the primary reason why I think the current war is far from over. A major problem that Thomas Edison could not solve was how to increase the transmission voltage of DC to transmit power over longer distances. So, back then, only customers within that few km from a DC power plant could get the power supply. With UHVDC, the transmission of direct current is making a comeback in the 21st century. This time, it will connect the globe.

Figure 2.16: UHV Technology to 1100kV

Source: ABB

Besides UHV technology, high temperature-superconducting transmission (HTS) is another technology that can make large scale GEI possible.[66] It has much lower losses than conventional power cables (25% ~50%) and has a very high transmission capacity – an 800kV HTS UHVDC line can transmit 16GW-80GW, which is about 2 to 10 times that of a current UHVDC today.[67]

Smart Grid, on the other hand, is like the brain of GEI. Its goal is to monitor and control the grids.[68] Traditionally, power monitoring and control are based on supervisory control and data acquisition/energy management systems (SCADA/EMA). It measures the voltage, frequency, and power status of the circuit breakers sent to the control centers. However, the problem is that this is not done in real time, so the data collected might have discrepancies. Smart Grid employs satellite navigation system and phase measurement unit (PMU) to synchronize measurements.[69] Wide area monitoring system (WAMs) is a cluster of PMU networks working together to improve the control and protection of grids.[70] With the integration of information and communication technologies (ICT), Smart Grid can run real time simulation and analysis with the data gathered to make decisions and give advice to maintenance strategies, maintenance plans, lower the operating costs, improve machine efficiencies, settlement of energy flows, etc.[71]

***

In my opinion, the key challenge of energy transition in the 21st century isn't just about energy generation, it is about meeting the global demand of energy in a clean and green way. It is about promoting sustainable energy worldwide. It is about designing a future with low-carbon, reliable and sustainable energy for our future generation. Our children's children, who haven't even been born yet, are counting on what we do today.

Chapter 3

The Wrath of Mother Nature

On 12nd December, 2015, under the United Nations, 195 nations got together in Paris for an agreement to fight climate change for the first time.[1] The central aim is to strengthen the global response to the threat of climate change and unleash actions and investment towards a low carbon, resilient and sustainable future. The goals of the Paris Agreement are simple: by keeping global temperature rise this century well below 2°C above pre-industrial levels and pursue efforts to limit the temperature increase even further to 1.5°C. The 1.5°C limit is a significantly safer defense line against the worst impacts of a changing climate.[2]

In 2018, the UN Intergovernmental Panel on Climate Change (IPCC) compiled the results of over 6,000 scientific studies and released a new report. Their find is that at the current rate, the global average temperature is likely to rise above 1.5°C as early as 2030.[3] This means we have a decade left to make drastic changes, or we will miss our target.

What was Temperature Rise in History?

But why should we care about 1.5°C and 2°C of warming? After all, temperature can vary by many degrees every day.

While the difference between 1.5°C and 2°C may sound very small for us, it is hard to imagine how half a degree increase in temperature will make an impact on the world. In reality, a one-degree global change is significant because it takes a vast amount of heat to heat up the ocean, the land, and our atmosphere by that much. In the past, a one to two-degree plunge was all it took to plunge the Earth into a little Ice Age.[4]

If you look back at the global temperature anomaly from 1880 until today, the average global temperature on Earth has increased by a little more than 1 degree Celsius since 1880. Two-thirds of global warming have occurred since 1975 at a rate of roughly 0.15°C to 0.2°C per decade. This rise in temperature is a direct result of population growth and carbon footprint left behind by human activities.[5]

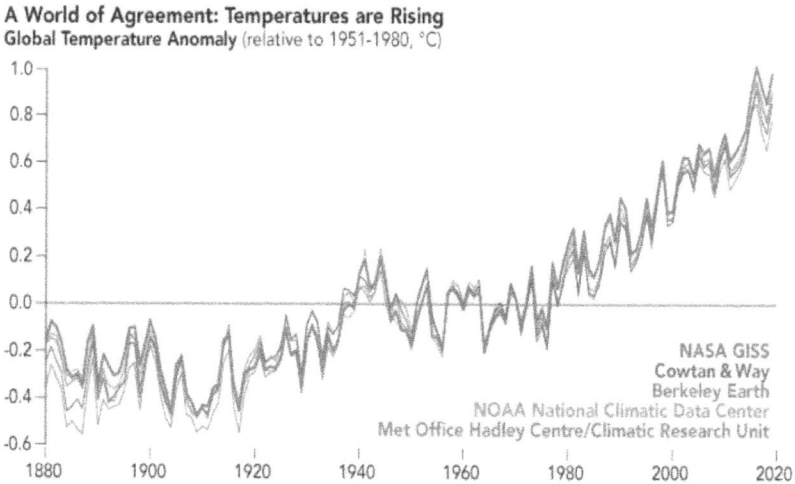

A World of Agreement: Temperatures are Rising
Global Temperature Anomaly (relative to 1951-1980, °C)

Figure 3.1: A world agreement: Temperatures are rising

Source: NASA's Earth Observatory

Today, by using the climate model, scientists can predict what that half a degree increase in temperature means for our future.

Arctic Ocean

The Arctic Ocean is like a refrigerator for the rest of the world.[6] It is particularly important because a change in the Arctic could affect the entire world's climate. Arctic ices help to cool the planet by the albedo effect.[7] To understand how it works, consider the sheets of ice in the Arctic working like insulation sheets by keeping the heat in the ocean from dissipating into the cold Arctic atmosphere. Melting glaciers and the ice in the Arctic Ocean means less sunlight will be reflected back to space. The loss of ice results in the Arctic Ocean absorbing the sun's energy, further raising the ocean's temperature. This disrupts seasonal patterns of heat exchange between the Arctic Ocean and the air.

So how does this affect our global temperature?

Remember that ocean current is critical to our weather. The flow of ocean current is responsible for regulating the global temperature by Atlantic Meridional Overturning Circulation (AMOC) - a large system of ocean currents.[8] It's like a conveyor belt, driven by differences in temperature and salt content that ensures the world's oceans are continually mixed, and that heat and energy are distributed around the earth. The key engine for the continuous flow of AMOC is the sinking of cold and salty water in the North Atlantic.[9] As salty water moves northward from the tropics, it cools off and becomes relatively denser than the surrounding water. This cold and salty water sinks to the bottom of the North Atlantic and begins to flow southward again along the ocean bottom. This causes more salty water from the tropics to flow northward, and the cycle continues. However, the melting of ice and glaciers in the North Atlantic and Greenland causes a large influx of freshwater and cap the global ocean circulation.[10] AMOC is one of the critical climate systems that global warming is pushing to a tipping point. Beyond that threshold, we will have an irreversible change that will cause catastrophic consequences for the planet.

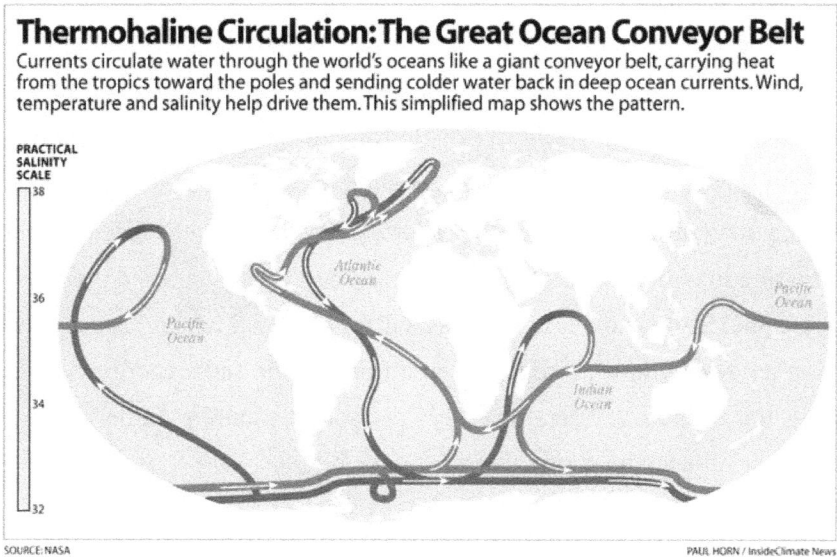

Figure 3.2: Atlantic Meridional Overturning Circulation (AMOC)

Source: NASA

If AMOC continues to weaken in the 21st century, we will feel it. It will mean much colder winters and much hotter summers. It will mean some parts of the world will have prolonged periods of time where there will be no rain. It will mean rising sea level in some parts of the world. It will impact the distribution of fish and cause a loss in biodiversity.

Deep Trouble of Rising Sea Level

In 2019, Venice suffered a week of catastrophic flooding. The second-highest tide was recorded, reaching a peak of 1.87 meters above the sea level on November 12, 2019. More than 85% of the city was flooded.[11]

Historically, tides high enough to flood Venice used to be relatively rare, occurring every two or three decades. Now, they've become increasingly regular – on the order of 5 years or less. While the great Venice flood in

2019 could be due to a combination of factors as well, the Venice mayor blames it on climate change.

Venice is just one of the examples how rising sea level is affecting coastal cities in the world. IPCC warned that the picture of Venice is what most coastal places will be like in the future.

Since the 1990s, satellites have measured acceleration in the rate of global sea level rise. A forecast from the National Oceanic and Atmospheric Administration (NOAA) shows that the global mean sea level is likely to rise at least one foot (i.e., 0.3 meters) above the 2000 level, even if greenhouse gas follows a low pathway from now to 2100. Professor Rahmstorf further confirms that the sea level may well exceed 1 meter (i.e., ~3 feet) by 2100, even if we continue the low emission scenario path.
12

But what does a meter increase in sea level mean?

According to the U.N. Atlas of the Oceans, eight of the ten largest cities in the world like Florida are just a few feet above sea level.[13] After a one foot increase in sea level, beaches will be regularly flooded at high tides. A two feet increase in sea level means a lot of roads would be submerged. A three feet increase in sea level means inland will be flooded when combined with a natural disaster like a hurricane. This means every single real estate across the coastal line will be devastated. If the sea level rose by as much as five feet, every port for offloading and loading goods would need to be replaced.

[Note: The sea level rising is accelerating by an average of 3.4 mm per year]

Possible future sea levels for different greenhouse gas pathways

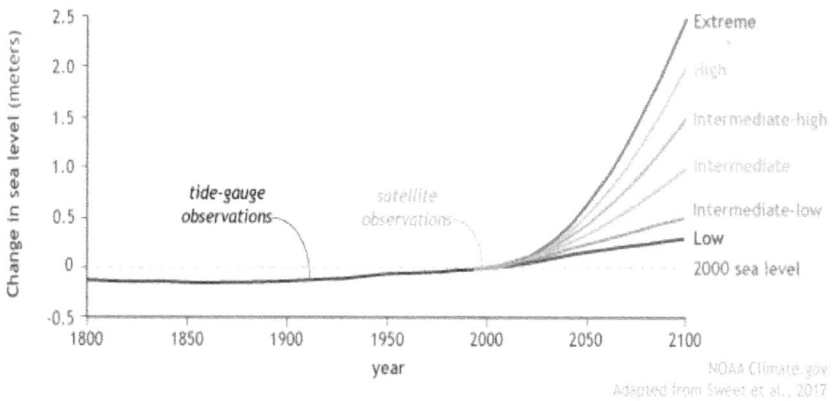

Figure 3.3 - Possible future sea level for different greenhouse gas pathways

Source: climate.gov

Source: NOAA Climate.gov graph, adapted from Figure 8 in Sweet et al., 2017.

So, how does global sea levels rise?

Global sea levels rise in two ways. First, glaciers and ice sheets worldwide are melting and adding water to the ocean. Second, the volume of water is expanding as the water warms. The amount of sea level attributed to the melting of ice was near twice the amount due to thermal expansion from 2005 to 2013. In other words, melting ice sheets are the major cause of rising sea levels. Antarctica and Greenland ice sheets contain more than 99% of the freshwater on Earth combined.[14] They are the two largest ice sheets in the world with a lot of sea level stored in them. Due to climate changes, the mass of these ice sheets, Greenland in particular, had ice loss accelerating at a rapid pace. Scientists predicted that if there were an increase in global temperature by 2°C to 3°C, all of Greenland's ice would be completely submerged in water. If all the ice on Greenland were to melt, it would raise global sea levels by 7.42m (24.34ft).

[Note: Together, Antarctica and Greenland contribute to 68% of all freshwater on Earth. If all the Greenland ice sheet melted, scientists estimate that the sea level would rise about 6 meters (20 feet). If all the Antarctic ice sheet melted, sea level would rise by about 60 meters (200 feet)[15].]

The Great Bushfire in Australia

In Australia, we just experienced some of the worst bushfires in Australian history. The fire spanned a period of six months – starting in September 2019 and ending in February 2020. Hundreds of fires have razed more than 12.6 million acres across different states. More than one billion animals had been killed.[16] 434 million tonnes of CO2 were emitted. Blood red skies hung over the states. Civilians were forced to wear masks to avoid breathing in hazardous haze.

Although Australia has always had bushfires, and the cause is due to a myriad of reasons, climate scientists had concluded that such an extreme fire season is at least 30% more likely because of global warming.[17]

Prior to the bushfires in Australia in 2019, Australia had suffered prolonged years of drought. In 2018, rainfall was very low across the southeast quarter of the Australian mainland (i.e., Sydney and Melbourne). The Australian government had implemented water restriction across the states. In South Australia, thousands of camels were shot dead as they were roaming the streets looking for water.[18]

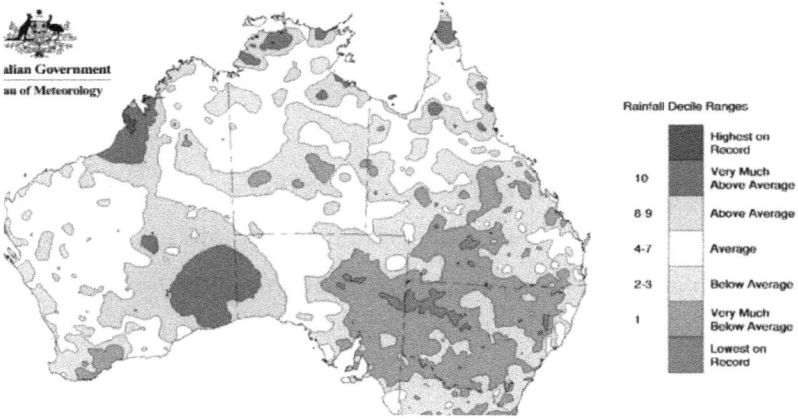

Figure 3.4 – Australia Rainfall Deciles in 2018
Source: Australia Government

Apart from that, 2018 was also the third-warmest year on record.[19] Many regions in Australia were in significant drought. Australia's national science research agency, the Commonwealth Scientific and Industrial Research Organization (CSIRO), projected that they have very high confidence that hot days will become hotter and extreme, and rainfall shortage will only become more intense.

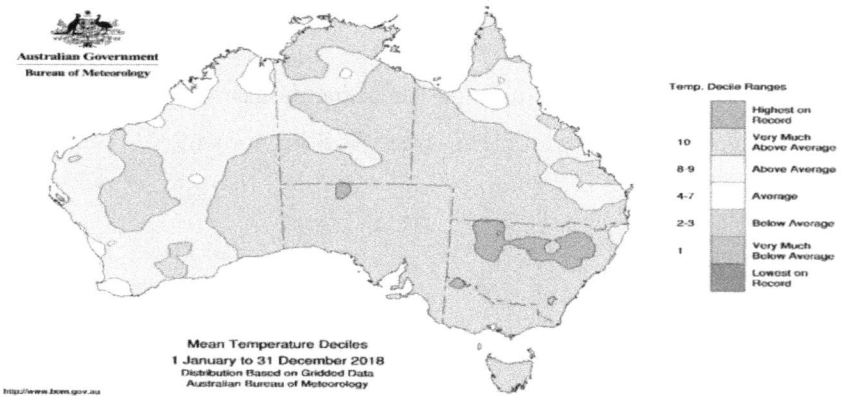

Figure 3.5 – Australia Temperature Deciles in 2018
Source: Australia Government

A Plastic Ocean

We are all surrounded by plastic. It is the single use of packaging we discard. It is cheap, lightweight, versatile, corrosion resistance, durable, and it is everywhere.

Since the invention of the world's first fully synthetic polymer by Leo Baekeland in 1907, plastic has become part of our daily life. The invention of this synthetic material has definitely raised our standard of living. Its unique properties encouraged technological innovation in the field of medicines, car manufacturing, pipe and cables manufacturing, etc. It may surprise you that the mass use of plastic didn't start until 1950 and has steadily increased ever since.[20] In just the first decade of the 21st century, more plastic was made than at any time in history. Almost all plastic is derived from materials made from fossil fuels. The process of extracting and transporting fossil fuels, then manufacturing plastic, contributes to creating billions of tons of greenhouse gases. According to the World Economic Forum, about 4% -8% of the annual global oil consumption is associated with plastic. If the world reliance on plastic continues, plastic alone will account for 20% of the oil consumption by 2050.[21]

The problem with plastic is that the largest amount of plastic waste comes from the packaging industry. Around 40% of plastic is used as packaging for single use.[22] Recycling is just not economical because of the high cost and low commercial value. Incinerating plastic will produce pollutants and release massive amounts of greenhouse gas emissions. Landfill them is not easy either. Plastic bags can take up to 10 to 100 years to degrade in landfills. Worst of all, plastic is so lightweight that it's often blown away from landfill sites into the ocean. Once it's in the ocean, plastic decomposes extremely slowly. It breaks down into tiny pieces known as microplastics that can be incredibly damaging to sea life. In fact, 80% of plastic in our oceans is from land sources. Ocean plastic pollution is a global tragedy for our oceans and sea life. Every day, approximately 8 million pieces of plastic pollution find their way into our oceans. Every year, 100,000 marine animals and sea turtles and one million sea birds are killed by marine plastic pollution.

In 1950, the world population of 2.5 billion produced about 1.5 million tons of plastic. By 2017, the 7 billion people on our planet produced over 320 million tons of plastic. This number is set to double by 2034.[23]

Today, studies estimated that there are about 51 trillion pieces of plastic in the world's ocean. Not one square mile of ocean surface on Earth is free of plastic pollution. At current rates, plastic is expected to outweigh all the fish in the sea by 2050.[24] If this problem is ignored, the foreseeable plastic ocean crisis will eventually come back to affect our future generation and us.

To prevent a plastic ocean, we must act together responsibly. We can start by stop littering. Stop using plastic bags. Encourage others to use secondhand items. Stop consuming and producing more than we need. Change our buying habits. And more importantly, educate our next generation about the importance of environmental conservation. After all, we want our children to inherit a future that is worth inheriting.

Remember, the ocean doesn't owe us. It gives. We take. But it can always take it back. If we continue to poison it, then don't expect it to feed us. If we keep on doing what we have been doing today, we won't survive. After all, the ocean doesn't need humans. Humans need the ocean.

The Future of Water

The oceans cover 71% of our planet. It is where 97% of the Earth's water is stored.

Ironically, for freshwater, the water we drink only accounts for less than 3%. Moreover, two-thirds of that is locked in ice caps or glaciers and is not accessible. So, the harsh reality is that we have less than 1% of freshwater drinkable. Most of this comes from rivers and freshwater lakes. To put it into perspective, if all the water on the planet would fill in a container with 100 liters, our usable supply of freshwater would be only about 0.003 liters (0.3% or one-half teaspoon).[25] This is how scarce our freshwater supply is for the increasing population.

Where Water is Found and the Percentage

Oceans	97.2%
Ice Caps/Glaciers	2.0%
Groundwater*	0.62%
Freshwater Lakes	0.009%
Inland seas/salt lakes	0.008%
Atmosphere	0.001%
Rivers	0.0001%
TOTAL	99.8381%

Figure 3.6: Where was is found and the Percentage

Source: www.usbr.gov

Right now, in 2020, more than two billion people, 26% of the global population, currently do not have access to safely managed drinking water.[26] But due to population growth, water users have increased six-fold and is rising about 1% every year. And by 2050, half the world's population may no longer have safe water.

We do have an emergency.

This is the future we will be heading if we do not take action today.

So, how is water being used today?

The largest use of water today is agriculture. It accounts for 70% of the world's freshwater use. Vast amounts of water are needed to raise livestock and for irrigation. To support the exponential rise in population growth, the amount of water used also increased accordingly. Because agriculture represents the most freshwater use, it is also the greatest sector to conserve water by changing irrigation practices. Techniques like drip irrigation can save up to 80% of freshwater than conventional irrigation, and some of these techniques even contribute to increased crop yield.[27]

For the remaining 30% of freshwater usage, on average, experts estimate that 20% is used in industry, and the household uses the remaining 10%.

In industry, water is needed to generate electricity, cool machines and refine minerals. Rainwater harvesting and wastewater recycling are some of the techniques used to reduce freshwater use.

For us, some simple ways we can help to conserve water is by reducing unnecessary water usage. Check for water leaks. Pick a more efficient washing machine. Install a low-flow showerhead. Although these are simple steps, every little bit helps. If everyone develops a good habit to conserve water and use it wisely, the result is that our future generation will also have the chance to enjoy a sufficient supply as we do.

In the later part of the book, I will discuss how technology is being used to solve the looming water crisis in the 21st century.

***

Unsustainable environment trends have finally caught up with us and are converging at a narrow window of time. This chapter is a snapshot of what is happening in our environment 20 years into the 21st century. Climate change is real and is a global concern.

Despite the obvious environmental predicament we are already heading, the GEO-6 (Sixth Global Environment Outlook) warns the overall condition of the global environment continues to deteriorate, driven mainly by population growth, urbanization, economic development, technological change and climate change. Under the current policy scenario, if this trend continues, goals like the Paris Agreement are unlikely to be achieved.[28] That is why the government, business and society should increase the scale and pace of environmental change proportionally. Only by addressing these environmental issues effectively and immediately, will we have a chance to steer the planet towards a more sustainable future.

Chapter 4

The Once and Future Money

Money does not grow on trees, but our modern banking system creates money far faster than trees can grow. When President Richard Nixon took the U.S. dollar off the gold standard in 1971[1],the Pandora's box of debts had opened, not just in the U.S but also in the world. The result is a level of prosperity the world had never seen.

However, the process changed the nature of our economy. Now, our dollar is experiencing a crisis. The global economy is in huge imbalance. Our children inherit trillions of dollars of debt.

This chapter is about the history of our dollar and its future in the 21st century.

Currency and Money

In my first book, Corruption of Real Money, I outlined the reasons why the dollar we earn today is not money.

Money should be a medium of exchange, a unit of account, divisible, durable, fungible, portable, and more importantly, it should be a store of value over a long period of time.

Currency, on the other hand, loses value every single year. It is a receipt of money but not money itself.

Under the gold standard, nearly all the countries fixed the value of their currencies to a specified amount of gold. Since a country can only have so much gold, this limits its ability to expand its money supply to account for all the goods and services in an economy. However, the constraint of gold was removed after the global dollar standard closed the gold window in 1971. In the absence of the gold standard, there is no way to protect savings from confiscation through inflation because there is no more safe store of value.

In fact, the primary reason why we are seeing real estate going up ten times, gold going from $35/oz to $1700/oz, is because governments around the around are racing to debase our currency supply through deficit spending. This gave the illusion that we have more currencies to spend to make us feel wealthier. But the truth is that expanding our currency supply dilutes everyone's purchasing power. After all, the currencies we use today are backed by nothing but a false promise of value.

To understand the current monetary chaos and what our future monetary system might be, it is important to understand the revolution of our international monetary system from the 20th century until today.

Revolution of the Monetary System

It may surprises you that roughly every 30 to 40 years, the world has a new monetary system.

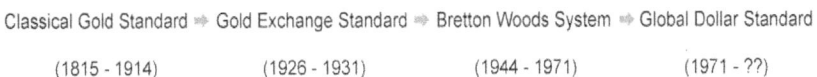

Classical Gold Standard ⇒ Gold Exchange Standard ⇒ Bretton Woods System ⇒ Global Dollar Standard

(1815 - 1914)   (1926 - 1931)   (1944 - 1971)   (1971 - ??)

Figure 4.1: Revolution of Monetary System

Source: Author

The classical Gold standard exists from 1815 to the outbreak of WWI.[2] Under a classical gold standard, all the currencies we use today were defined as a definite weight of gold. For example, the dollar, is defined as 1/20 of an ounce of gold. The British Pound, on the other hand, is ¼ of an ounce of gold, etc. This implies the exchange rates between national currencies were fixed and were not meant to be set by the governments.

One of the primary reasons why the U.S. prospers is because the U.S. dollar was pegged to sound money – gold. The government did not pick gold to be a monetary standard. The free market picks it as the best money because of the stability it creates in the monetary system. The demand and supply of gold were set by the free market, not by the government's printing press. One of the key benefits of the classical gold standard is that it keeps the government's inflation in check. It keeps the balance of payment of countries always in balance. That was why the price of things back then was relatively stable under the classical gold standard.

However, if the classical gold standard went so well, why did it break down?

The classical gold standard broke down because the governments around the world indebted themselves to fund the war during WWI. Eventually, the world reached a point where there was not enough gold to cover all the currency created during the wartime period. For that reason, the world needed to redefine the price of gold for all currencies like the *franc, mark, pound, dollar,* etc.

Back then, Britain was the financial center of the world. The British pound had been traditionally defined at a weight of ¼ of an ounce of gold or £ 4.24/oz. By the end of WWI, £4.24/oz had become far too high. The foreign exchange market pushes down the price of the pound to £3.5/oz to be competitive in the international markets. However, to bring it back to the Classical Gold Standard of £4.24/oz, Britain would have to deflate its currency, which was politically impossible. Britain was caught in a situation where she had to choose between saving her export industry or worsening unemployment. Realizing that going back to the Classical Gold Standard was impossible, Britain sought help from the U.S., which later on led to the birth of the next monetary system – The Gold-Exchange Standard.

After WWI, the U.S. held most of the gold in the world. In the post WWI era, many U.S. dollars were held outside the U.S. by allied nations. To help Britain, the U.S. chose to inflate its own currency supply to maintain a healthy exchange rate for British export. Since the British Pound was the world's reserve currency at that time, and the USD was pegged to most of the Gold, this formed a relationship like below between gold, the USD, the British Pound and the world's currency.

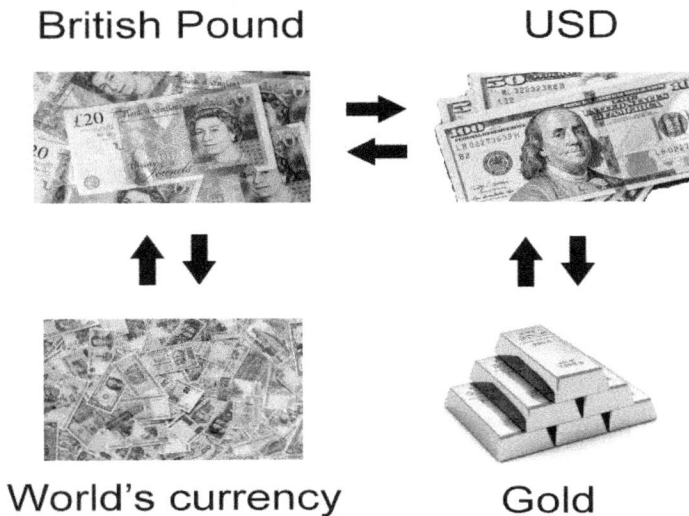

Figure 4.2: Gold Exchange Standard (1926 – 1931)

Source: Author

In the end, Britain never really went back to the *Classical Gold Standard*. It chose to use cheap credit, with the help of the U.S., to stabilize its economy.

The *Gold Exchange Standard* was a man-made system. It relied on the U.S. and Britain's central bank monetary policies to keep its economy stable. So, when Britain inflates their currency supply, she will have a deficit in the Balance of Payment.

But, like any other man-made fiat currency system, the *Gold Exchange Standard* is intrinsically flawed. A lengthy deficit spending and the loss of confidence in the monetary system eventually caused the *Gold Exchange Standard* to collapse like a house of cards. Countries around the world competed with one another by devaluation. That led to a fluctuation in exchange rates. International trade went into a standstill.

From 1931 to 1945, the world suffered a global Depression and WWII. From 1933-1934, the U.S. went off the gold standard to get out of the Great Depression. Citizens were forbidden to own any gold at all. After 1934, the U.S. restored a gold standard, which pegged gold at \$35/oz. By supplying the Allied nations with weapons, supplies, and other goods during WWII, and collecting much of its payment in gold, the United States owned the majority of all the gold in the world by the end of the war.

To restore international monetary order, in July 1944, the U.S. gathered 44 Allied nations at the Mount Washington Hotel in Bretton Woods, New Hampshire, for a conference. The purpose was to set up a system of rules and procedures to regulate the international financial system after WWII, which was referred to as The Bretton Wood System. Two international agencies, the *International Monetary Fund* (IMF)[3] and its sister organization, *Bank for Reconstruction and Development* [4], which later known as *World Bank*, were created at the conference.

The *World Bank* was designed to make loans to developing countries to help them build a stronger economy. The IMF, on the other hand, was to promote monetary cooperation between nations by maintaining a fixed exchange rate between countries.

After the war, the entire international financial system was devastated. There was a huge imbalance in the quantity of currencies between countries. Because most of the international trades were based on borrowing from the U.S., she was the biggest creditor nation. During that time, many countries around the world were holding U.S. dollars, and the U.S. was the biggest gold holder in the world. That was how it was agreed that the U.S. dollar would become the world's reserve currency backed by gold at $35/oz. That is why we often heard the U.S. dollar was as good as gold. This was known as *The Bretton Woods System* (1944-1971).[5] It requires all the world's currencies to exchange the U.S. dollar before they could claim Gold. Ever since, the world has become dependent on the U.S. dollar.

While the U.S. economy was enjoying its financial fortune, the U.S. government began to diverge away from production. It chose to invest in military spending. U.S. President Lyndon Johnson's Guns and Butter Administration was a classic example of that. The country was led to fight the Vietnam war. They competed with the U.S.S.R in the space program by sending men to the moon at the expense of its unshakable financial position after WWII.

With this huge deficit spending in the 60s, the U.S. was on the brink of losing her financial fortune. That was why she needed to find a way to keep her gold at all cost. On August 15, 1971, U.S. President Nixon convinced the entire world to replace gold with the U.S. dollar by closing the gold window. When that happened, all the world currencies became fiat simultaneously. This is why we now see gold going up from $35/oz to $1700/oz, and real estate going up ten times today.

In the absence of a *Gold Standard*, there is no way real adjustment mechanism can keep trade imbalance in check. With the U.S. dollar as the world reserve currency, and the Federal Reserve's ability to expand and contract the currency supply, the U.S. has the privilege to print unlimited U.S. dollars to exchange all the goods and services produced by the rest of the world.

Ironically, thirty years after the collapse of the *Bretton Woods System*, the world enjoyed an unprecedented boom in history. Under the new *Global Dollar Standard*, the Federal Reserve no longer requires to hold gold to back the currency created. There is absolutely no constraint on how much currency the Federal Reserve can create. Since our currency is pegged to the U.S. dollar, the nature of our currency has changed as well.

Below is a chart showing the total credit market debt in the U.S. It is the total amount of debts (credits) in the entire U.S. economy. In Q3 1964, the total credit in the U.S. hit 1 trillion for the first time in history. Over the next 55 years, it has expanded 75 times. This explosion of credit has changed the nature of our economy.

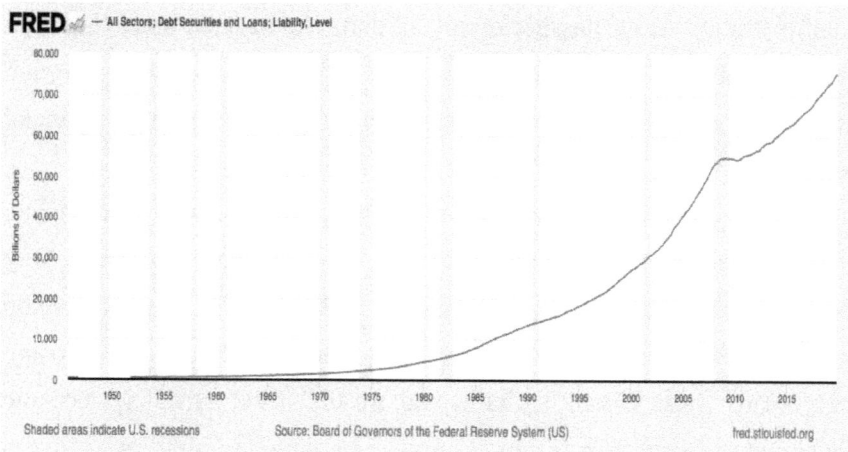

Figure 4.3: TCMD in the U.S.

Source: Federal Reserve

In the past, our economy grows by production and saving. Today, our economy grows by credit creation and consumption. In order for the economy to grow, the world must continue to borrow and incur in more debt. The proliferation of credit creation around the world dilutes the dollar everyone holds. It is the primary reason why our living standard will continue to decline. Savers become losers. Debtors become winners. Future generation, on average, will only find their standard of living continue to decline. This is the legacy they are inheriting from us – the legacy of debt.

Trade Imbalance Under the Dollar Standard

With the revolution of the international monetary system, a change in the nature of money means a change in international trade. In fact, international trade is the reason that leads to a revolution in the international monetary system.

Under the Classical Gold Standard, international trade was settled in gold. Surplus countries experienced inflation. Deficit countries experienced deflation. Business cycles help trade return to balance as a result. It is a self-adjusted mechanism by the free market using sound money.

Under the Bretton Woods system, trades were settled in dollars pegged to gold at $35/oz. Currency values didn't move up and down against each other. The exchange rates were all fixed. The Bretton Woods System was designed to replicate the self-adjustment mechanism of gold. Unlike the Classical Gold Standard, the Bretton Woods system has no shortage of dollars. It only has a shortage of gold! When the U.S. runs large trade deficits, and as other countries began to exchange the receipt of money (i.e., the USD) to gold, the U.S. did not have enough gold. That was why President Nixon had to end the Bretton Woods System and close the gold window in 1971.

Without the Bretton Woods System, there is no way to agree on how international trade works. Because there are no substitutes, the world had no choice but to continue to accept the USD so that the carousel of the international financial system could continue. The Global Dollar Standard was born under such an arrangement. Under this new monetary system, governments around the world can print as much currency as they want. Some currencies started to float against one another because of the central banks' currency manipulation. Worst of all, trade between countries is no longer balanced.

The chart below is the balance of payment of the U.S. You can see how the U.S. is running huge trade deficits with the rest of the world. This set off a global economic boom never seen before in the world. The transition to the Global Dollar Standard created an illusion to the rest of the world that we are living in a new era.

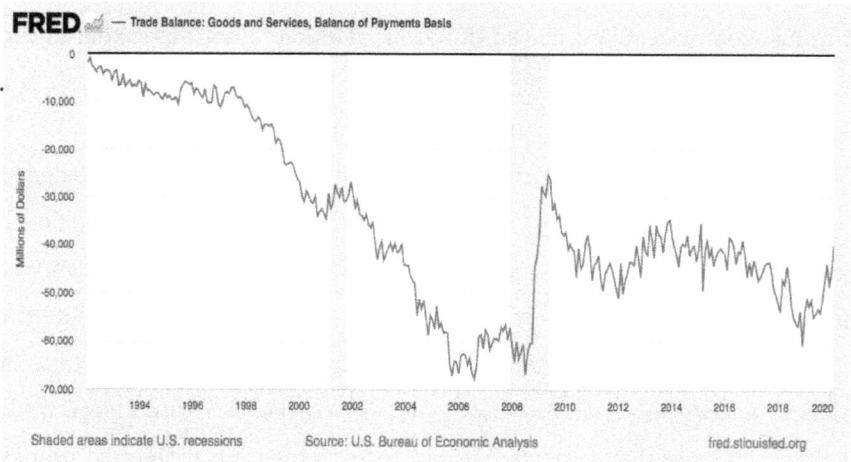

Figure 4.4: Trade Balance in U.S.

Source: Federal Reserve

Under the Global Dollar Standard, trade surplus countries like China don't want their currencies to appreciate. China wants to devaluate its currency to favor its export-led growth economic model. So, as the U.S. runs huge trade deficit with the rest of the world, China has been running a huge trade surplus with the U.S. To do so, the People's Bank of China (PBOC) prints their own money to buy USD or US denominated assets like the U.S. Treasury Bond to prevent its currency from rising. This keeps the dollar strong and the RMB weak. This currency manipulation has been fueling China's economic growth even today.

While the U.S. trade deficit is the driver for global economic growth, it is only possible because many people in China and third world countries are willing to work for less than 10 USD a day.[6] However, many Chinese factory workers cannot afford to buy the product they make. While the U.S. is flooding the world with dollars, the growing population in China and many third world countries are flooding the world with cheap labor. Driven by cheap labor and profits, companies in the West deindustrialize themselves by moving their manufacturing base to developed countries, leaving only service sector jobs onshore. That is the main reason why wages in many developed countries have stagnated for decades.

Right now, manufacturing companies in the West are trying to bring the manufacturing base back to their country. However, after decades of imbalance, the West is now caught in a position because their wages are too high compared to factory workers in third world countries. Bringing back manufacturing jobs becomes uneconomical as workers in the West demand higher wages, which translates to a higher price of products. And if workers in the West accept lower wages, they cannot afford to pay the bills.

This is the result of the Global Dollar Standard the world is experiencing right now.

This is why a new monetary system will likely happen in the 21st century to correct the imbalances. And I believe that when combined with technology, gold, the once future money, has an important role to play.

PART II

Crystal Balls of the 21st Century

Chapter 5
Technological Singularity

Technological singularity describes a hypothetical future point in time at which technological growth becomes uncontrollable, irreversible, and unforeseeable to human civilization. It happens when the pace of technology changes so rapidly that our life will be irreversibly transformed.

By definition, technological singularity is something unpredictable. It is just like a man living from 1850-1930 couldn't have imagined the drastic technological change like Wi-Fi, genetic engineering, cloud computing, artificial intelligence, nanotechnology, quantum computing and every other innovation we have today. If you fully understand the significance of singularity, you will realize that the future is not what we think it's going to be.

Law of Accelerating Return

Ray Kurzweil, the author of *The Singularity is Near*, states that the fundamental measure of information follows exponential trajectories.[1] It is contrary to our conventional "intuitive linear" view.

But what is exponential?

Let's do a thought experiment together.

Imagine we have a magic eyedropper that drops water onto your hand. What makes this eyedropper magical is that the volume of water it drops doubles every minute. (i.e., for the first minute, you have one drop of water; for the second minute, you have two drops of water, etc.)

Now, imagine if we bring this magical eyedropper to the highest point in Santiago, Bernabeu Stadium. Starting at 12:00 pm, we drop one drop of magical water to the center of the stadium every minute. How long do we have to escape before the entire stadium is flooded with water?

[Note: Santiago Bernabeu Stadium has a field size of 105m x 68m. Height is 45m. Assume a drop of water is 0.05mL].

Figure 5.1: Santiago Bernaberu Stadium

Source: Author

The answer is that you have 43 minutes to escape the stadium.

Perhaps you might be overestimating or underestimating the time.

But this is not the real point I want to illustrate.

The real question is: What time would the stadium still be 97% empty?

The answer is 12:38 pm.

Five minutes before the Santiago Bernabeu Stadium is flooded, it is still 97% empty.

This is the nature of exponential.

At the beginning, the trend of exponential is so flat that it looks like there is no trend at all. Then the rate of change accelerated so quickly that it becomes unpredictable.

If technological growth is that magical drop of water, this implies we won't experience a century of progress in the 21st century. Most likely, we will experience 20,000 years of growth at today's rate of growth.

Moore's law states that the number of transistor count doubles every 18 months.[2] This has been true in the past fifty years. And that is the reason our computers get more and more processing power every year. One of the key challenges of future engineering of nanoscale transistors is the design of gates. In fact, industrial leader Intel just announced it would delay its 7nm process node for at least six months.[3] Moving from 7nm to 5nm is going to be more and more difficult because of electrical instability.

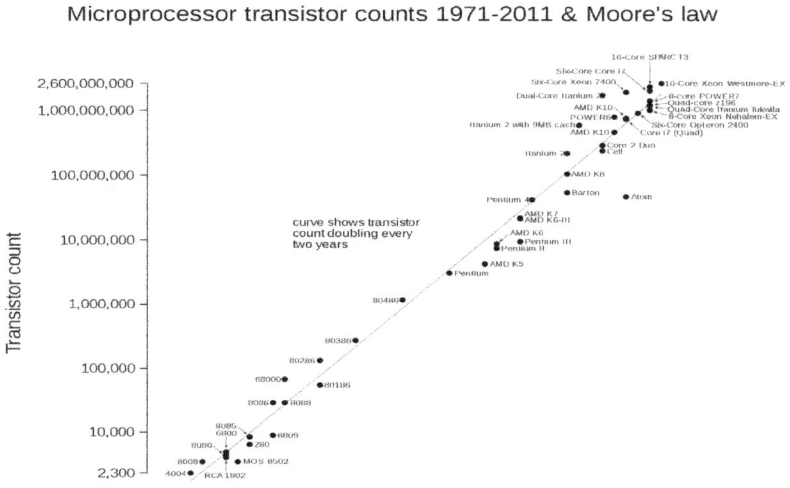

Figure 5.2: Moore's Law

Source: Wgsimon wiki

Clearly, Moore's law may not be able to go on indefinitely. Computer scientists need to devise more efficient algorithms to help conventional computers keep pace.

But does that means the accelerating growth of technology will peak with Moore's law?

On 23rd October 2019, scientists at Google announced that they had achieved quantum supremacy – a long-awaited milestone in quantum computing.[4] Quantum computing carries out specific calculations that are beyond the practical capabilities of conventional computers today. Google estimates that the same calculation run on a quantum computer would take even the best classical supercomputer 10,000 years to complete.[5]

A new law called Neven's law states that quantum computers are gaining computational power at a doubly exponentially rate. Doubly exponential growth is so singular that it is hard to find examples in the real world. The rate of progress in quantum computing may be the first.[6]

[Note: Exponential growth is fast. It means something grows by the power of 2. Doubly exponential growth is more dramatic. It means quantities grow by powers of powers of 2.]

Snapshot of the 21ˢᵗ Century

How is technological singularity going to help us with the growing energy and environmental issues in the 21ˢᵗ century? How is it going to shape our future? How is it going to shape us as a human race?

We are living in a very interesting time in history:

- The human population doubles every twelve years.

- Oil, the energy source we have taken for granted, has reserves of less than 47 years.

- Global energy policymakers respond by shifting from concentrated energy sources to a less concentrated energy source to reduce carbon footprint.

- We are close to a tipping point where the effect of global warming becomes irreversible.

- Overcapacity and overproduction in China have flooded the world with cheap manufacturing goods.

- We are living in a world of abundance and waste.

The world is trying to seek endless economic growth through GDP. However, our energy system and environment can no longer support that type of growth. Countries are rapidly depleting the world's scarce resources to produce more than we can consume. Above all, the end result is a mountain of waste, causing irreparable damage to our environment, resulting in economic loss.

Aren't we living in a contradictory world in the 21st century?

In the chapters ahead, I will talk about several key, measurable technological trends that are shaping our future in the 21st century. From computing to medicine, miniaturization to nanotechnology, creating life to space exploration, technology is going to transform our lives in an unimaginable way. It will solve a lot of problems we face today, but it will create new problems at the same time. And I believe the worst thing we can choose in this exciting time is to be technologically illiterate.

I believe there are good reasons to believe we are in a turning point in history. Within the next two decades, we might see a meaningful understanding of how our brain works. There is a high possibility that the brain-computer interface will become mainstream in the near future.[7] The combination of human intelligence, computer processing power and memory will expand our potential. The intersection between genetics, nanotechnology, and robotics will mark the singularity of technology. It could be entirely possible that we can reprogram our cells using computers to eliminate disease[8], aging, or even cheat death by uploading our consciousness in the cloud while controlling matters in the physical world.

I believe the reason you picked up this book is because you want to know what the future holds. The rest of the book is designed to help you familiarize yourself with several key technologies. Please be advised there will be a lot of history of technology to follow. And the further you look into the past, the further you are likely to see in the future.

Chapter 6
The Birth of the Computer

In the historical drama film, The Imitation Game, German armed forces send messages securely by using a type of enciphering machine called Enigma.[1] Hitler and his German Army used it to encrypt detailed situation reports at the battlefronts and minutiae of war.

To win the war, the British needed to intercept the messages. The faster they could break the messages, the faster they could contain the situation. But, at that time, the Enigma cipher machine was known to be unbreakable.[2] In the movie, a Cambridge mathematician, Alan Turing, set out to defy the odds. His mission objective was to design a machine that could decrypt the Enigma message.

Turing was one of the fathers who contributed to inventing the modern computer. In 1937, he published a paper titled *"On Computable Numbers, with an application to the Entscheidungs problem,"* which became the foundation of computer science.[3] In his paper, Turning presented a theoretical machine that could solve any problem described by simple instruction encoded on a paper tape.

He called it the Turing machine, which is the original idealized model of a computer.

Figure 6.1 : The Turing Machine

Source:rutherfordjournal.org

Who invented the Computer?

However, Turing was not the only person who invented a computer. You might be surprised that the history of computer actually dates back to the 1800s.

Ada Lovelace, a British female mathematician, is the world's first programmer.[4] Lovelace wrote the world's first machine algorithm for early computing on paper to calculate some rather complicated numbers – *Bernoulli numbers*.[5] This was the first time in history that computers have done complicated things.

As a teenager, Lovelace met Cambridge mathematics professor Charles Babbage, who had invented an automatic mechanical computer to tabulate polynomial functions without error. Babbage called it the Difference Engine.[6] Even so, Babbage never built the Difference Engine because of financial strain. Instead, he designed the Analytical Engine[7], which uses punch cards for input and output (I/O).

[Note: A software language developed by the U.S. Department of Defense was named "Ada" in her honor in 1979.[8]]

Figure 6.2: Victorian-era computer called the Analytical Engine designed by Charles Babbage

Source: Science Museum London

Unlike today, where computers are all electronic-based, in the early days, mechanical computers used levers and gears. They could be analogue or digital. For analogue models, they used curved plates or slide rules for computation, and digital ones were built using gears.

The Chronicle of Computer Memory and Storage

So, how does a Turing machine work? How does it have memory? How does it compute?

Imagine a Turing machine called Newbie. Newbie consists of an infinite length tape, which acts as the memory of a computer. Each tape is divided into cells. In each cell, there is either a 1 or 0 or empty space. Above one cell of the tape is a Newbie's head, which can move either left or right and can read the symbol in the cell below. Apart from reading, Newbie's head has different functions as well. It can erase symbols and write new ones into the cells.

Let me illustrate with a simple program in the diagram below. Let's say we have a long string of 0 or 1 in a tape. According to state 5 of the program, if the Newbie's head is above a cell, which has number 1, it will erase it and write 0, move left, and go to another state, say state 7. This is how the tape was being used to read and write a program.

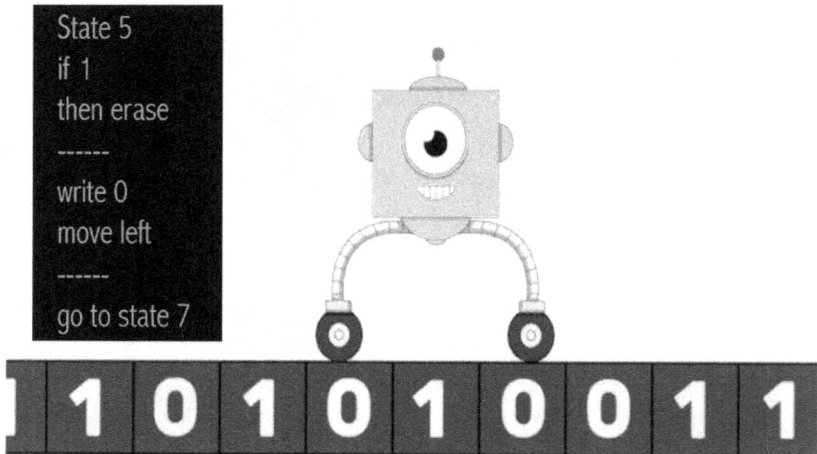

Figure 6.3: Reading and Writing program with the Turing machine

Source: Author

Say, we have "1 1 0" printed on the tape, and we want to convert it into "0 0 1." This can be done by passing the following instruction in the table below:

State	Symbol Read	Write Instruction	Move Instruction
0	Blank	None	None
	0	Write 1	Move tape to the right

	1	Write 0	Move tape to the right

Table 6.1 State Table of inversion program

Source: Author

So, when Newbie's head reads 0 in a cell, it rewrites the cell with 1 and moves the tape to the right. On the other hand, if the head reads 1 in a cell, it replaces the cell with 0 and moves to the right.

Pretty simple, right?

Even so, a simple program like this is incomplete because it doesn't know when to end.

To end the program, a stop state is needed.

When the head reads a blank cell, it should be directed to a stop state to end the program.

State	Symbol Read	Write Instruction	Move Instruction	State
0	Blank	None	None	*Stop State*
	0	Write 1	Move tape to the right	*State 0*
	1	Write 0	Move tape to the right	*State 0*

Table 6.2 State Table of inversion program

Source: Author

With more states, the Turning machine can perform more complex functions other than simple inversion. That was the computer blueprint that Turning came up with. And later on, this led to the invention of modern digital computer, which did not happen until 1937.

The earliest computer memory is a paper punch card.[9] By 1940, punch cards had largely standardized into a grid of 80 columns and 12 rows. It allowed a maximum of 960 bits of data to be stored on a single card.[10] Back then, data was input by punching holes on each column in a punched card. Once a punched card was completed, a return key was pressed to store that information. A device called a punch card reader would read these data from these punched cards. The standard measure of reading speed was cards per minutes, which was around 300 to 2000 CPM.[11] And the 80 columns in the punch cards translated to reading 250 characters per seconds.

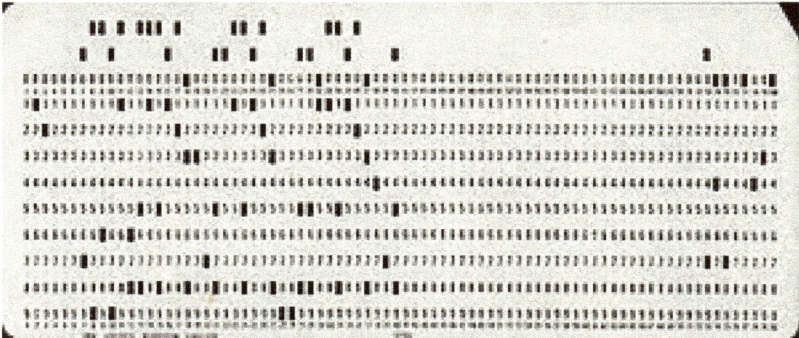

Figure 6.4: Punch Card

Source: wikipedia

The largest program ever punched on punch cards was the U.S. Military's Semi-Automatic Ground Environment or SAGE[12], an Air Defense System that became operational in 1958. The main program was stored on 62,500 punch cards, which is roughly equivalent to 5 MB of data.

Punch cards were useful and popular storage for decades. They don't need power. Paper was cheap and reliable. However, punch cards were slow. You can only write it once. Once written, the data will be stored permanently. You cannot easily un-punch the hole afterwards. Imagine you are writing a program; you will need a stack of these punch cards.

Since each punch card can only hold 80 characters, how big do you think Google's 15 Exabyte data warehouse would be if we were still using punch cards to store data today? So, a faster, more flexible computer memory was needed.

An early substitute for punch cards was Delay Line Memory, invented by an American engineer and computer pioneer, John Adam. Presper Eckert Jr.[13]

Here is how it works. Imagine you have a tube filled with liquid. You put a speaker at one end and a microphone at the other end. When you pulse the speaker, it generates a pressure wave. It takes time for the wave to propagate to the microphone, which converts back into an electrical signal using an amplifier. This propagation delay can be used to store data. After working at ENIAC, Eckert and his colleague, John Mauchly, went to build a better computer called EDVAC[14], the earliest electronic computer, using Delay Line Memory. EDVAC has 128 delay lines and is capable of storing 352 bits.[15] This allows EDVAC to be one of the earliest computers to store programs.

The problem with Delay Line Memory in EDVAC is that you can only read one bit of data from a tube at any given time. If you want to access a specific bit, you have to wait for that bit to come around in the loop. It is a sequential access memory. You cannot access any bit at any time. It is not Random Access Memory (RAM) or volatile memory.

By the mid-1950s, Delay Line Memory was largely obsolete and replaced by magnetic core memory. To visualize how magnetic core memory works, imagine you have a little magnetic donut, called a core, which is looped by a current-carrying wire. When you run an electrical current through the wire, it will magnetize the core. If you turn the current off, the core will stay magnetized. And if you reverse the polarity of the current, the magnetization direction changes accordingly.[16] This mechanism allows 1 and 0 to be stored. If you have an array of these 1,024 cores arranged into a 32 by 32 grid, 1024 bits can be stored. By taking one step further, and you have sixteen arrays stacked on top of each other, you will have 16 Kilobits of storage. This is called core memory, where any bit can be accessed at any time.

Back then, computer storage started at around $1 per bit. By 1970, the cost fell to 1 cent per bit. Consider a photo today is around 5 Mb, which is 4,000,000 bits. Would you pay $40,000 USD today just to store a photo?

In 1951, Eckert and his colleague Mauchly started their own company and designed their own computer called UNIVAC – one of the earliest commercial sold computers.[17] Instead of using magnetic cores, magnetic tape was used. It is a long, thin, flexible strap of magnetic strip material stored in reels. If you are old enough, you've probably seen them in cassette tapes. The strap can be wound back and forth inside a tape drive. Inside the tape drive is a write head, which is like a magnetic core with current-carrying wire. When the current passes through it, it generates a magnetic field that magnetizes a small section of the tape underneath the write head. The change in polarity of the current is used to store 0 and 1 on the magnetic tape. AUNIVAC used a half-inch (12.7mm) wide tape parallel data track, each one able to store 128 bits of data per inch or megabytes of data per reel. But tape is sequential in nature, and you would need to rewind or fast forward hundreds of feet of tape for a long time to get to the data you wanted. It is definitely time-consuming. This is when magnetic drum memory comes in.

A magnetic drum was basically a metal cylinder coated with magnetic material to record data. It rotates continuously around thousands of revolutions per minute. Positioned along its length were dozens of read and write heads. These heads would wait for the right spot to rotate underneath them to read or write bits of data. This technology directly leads to the development of Hard Disk Drives. In fact, IBM RAMAC 305 was the first computer with a disk drive, which has a storage capacity of roughly 5 MB.[18] By 1970, the hard disk drive had become commonplace.

Figure 6.5: IBM RAMAC 305

Source: www.snopes.com

A close cousin to the disk drive is probably the Floppy Disk (FDD). You might even recognize it as a save icon in some of your applications. The FDD was invented at IBM by Alan Shugart in 1967.[19] The 5.25-inch disks were dubbed "floppy" because the diskette packaging was a very flexible plastic envelope. It was ubiquitous from the mid-1970s up to the mid-90s. It uses magnetic medium. The 5 ¼" was first released in 1976, capable of storing 95.8KB of data. By 1980, 3 ½" floppy disk was used with an initial capacity of 360KB of data. In the 1987 version, an HD version was released that could store 1.44MB of data. However, a decade later, the FDD went out of favor.

Figure 6.6: Floppy Disk

Source: Author

Optical storage came about in 1972. More popular storage such as Compact Disk and DVD took off in the 1990s. Intrinsically, these technologies are very similar to the FDD and Hard Disk. But instead of storing data magnetically, optical disks have little divots on the surface that caused light to be reflected differently, which is then captured by an optical sensor and encoded into 1 and 0.

Today, solid-state technologies are replacing optical disks. We are moving computer storage with no moving parts, like USB and Solid State Drives (SSD). SSD access times are typically 1/1000th of a second.

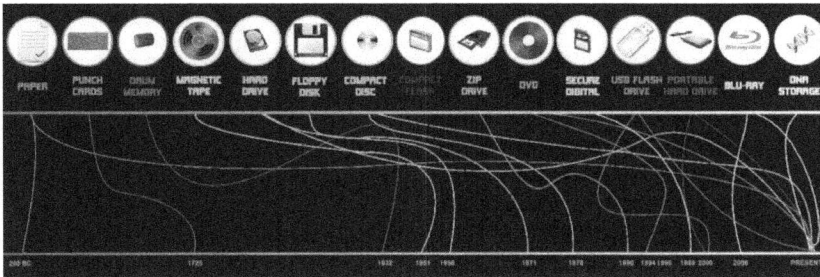

Figure 6.7: Computer Storage

Source: datasolutionslabs.com

We have come a long way in computer storage since the 1940s. Computer storage has followed an exponential trend, where price has come down with more power. From early core memory costing millions of dollars per megabyte, we have fallen to mere cents by 2000. Today, the cost is only a fraction of a cent.

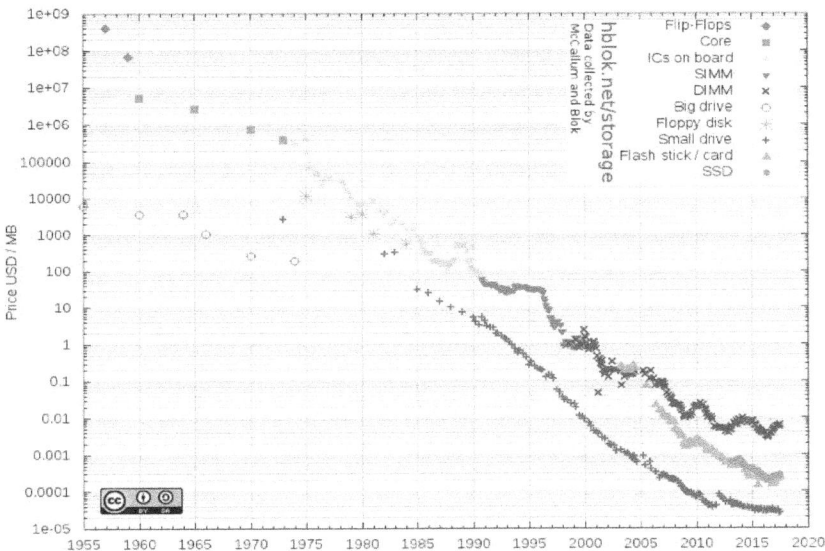

Figure 6.8: Cost of Computer Storage

Source: hblok.net/blog/storage/

More on Computer Memory

Computers today have two types of memories – primary memory and secondary memory. Primary memory is the memory of a computer directly accessed by the processor. They are volatile and require power to maintain the stored information. However, if the power is interrupted, the stored data will be lost. Secondary memory, on the other hand, refers to the storage devices we talked about earlier, such as SSD, USB, CDs and DVDs. They are much slower but have greater storage capacity. For example, a computer might have 16 gigabytes of RAM but one terabyte of hard drive storage. Also, secondary memory is non-volatile. Information in the secondary memory is still retained, even if power is lost. Unlike primary memory, the CPU does not directly access secondary memory. Data from the secondary memory is first loaded into the RAM before sending it to the processor. By loading programs and files into primary memory, computers can process data much faster.

Two of the main examples of primary memory are perhaps RAM (Read Access Memory) and ROM (Read Only Memory). Think of RAM as the short-term memory of a computer. It is temporary. It stores and read data at any time from any location.

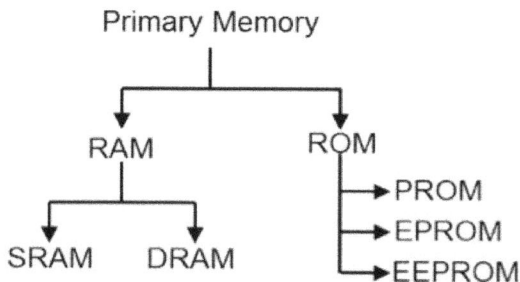

Figure 6.9: Cost of Computer Storage

Source: Author

In general, RAM can be further broken down into SRAM (Static RAM) and DRAM (Dynamic RAM). SRAM is the fastest among all memory. It is volatile and can retain memory using flip-flops as long as the computer is powered. SRAM is used for computer's cache memory. The architecture of SRAM uses several transistors in a circuit to store one bit. DRAM, on the other hand, uses capacitors and transistors per bit.

The most common type of DRAM used in modern processors is perhaps synchronous dynamic random access memory (SDRAM).

[Note: flip-flops or latch is a circuit that has two stable states and can be used to store information.]

Fig 6.10: SDRAM

Source: hblok.net/blog/storage/

[Note: If you have an 8 module RAM, open it up and zoom into one of the modules. The first thing you will see are 32 squares of memory. Zoom into one of those squares, and you can see each square comprised of 4 smaller blocks. If you zoom in again, you will see the matrix of each individual bit.]

Back in the 1970s, when DRAM was first introduced, it was asynchronous with the CPU's clock. This posed a problem because the CPU does not know the exact timing at which the data will be available from the RAM on the input-output bus. The CPU needs to wait between memory access. Also, DRAM will not work as fast as the processor, which gets better and better every year. SDRAM fixes this problem by synchronizing the I/O clock frequency with the RAM internal clock frequency.

DDR SDRAM (Double Data Rate Synchronous Dynamic Random-Access Memory) is a type of SDRAM that fetches data on both the leading edge and the falling edge of the clock signal that regulates it.

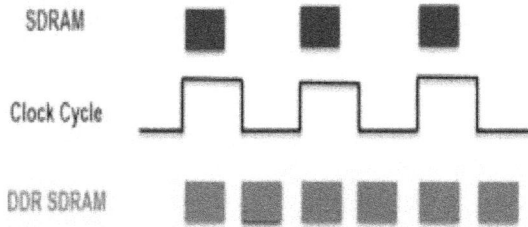

Fig 6.11: SDRAM and DDR SDRAM

Source: Author

ROM, on the other hand, is the long-term internal memory. Unlike RAM, ROM is non-volatile and can retain data without the flow of electricity. It is an essential chip that permanently writes data or programs. There are three types of ROM: PROM, EPROM and EEPROM. PROM (Programmable Read Only Memory) is one time programmable only. EPROM (Erasable Programmable Read Only Memory) and EEPROM (Electrically Erasable Programmable Read Only Memory) are reprogrammable. The main difference between EPROM and EEPROM is that EPROM is erased by UV rays, and EEPROM is erased by electric signal.

Even with the drastic improvement in the speed of memory over the years, the market is always going to demand a faster, efficient, and cost-effective memory solution. They will always look for the next generation memory that can support cutting edge technology in the 21st century.

The Future of Computer Memory

In-memory computing (IMC) is a type of middleware software that allows data storage in RAM across a cluster of computers and processing it in parallel. It focuses on real time insight, speed and scalability.

The reason that IMC exists is that the traditional access time of data storage is slow. As data grew over time, the time required to access data and perform analytics also increased. Also, when businesses seek quicker access of data and analytics to make business decisions, IMC can provide superfast performance to meet that demand. It gives business real time insight so that they can take immediate actions and responses.

IMC can deliver faster speed because the data is stored in a distributed fashion. The storage of data occurs in RAM across multiple computers instead of a centralized database. In contrast to a single, centralized server managing and fetching the data requested by queries, IMC's distributed storage fashion allows many different computers across different locations to instantly share and access data, making scaling possible.

Apart from business usage, as the Internet continues to expand, connecting people and devices like fridges, thermostats, light bulbs, jet engines and even heart rate monitors, IMC can produce streams of information in real time that will not just inform us but also protect us, make us healthier and help us live richer lives.

The Future of Computer Storage

In the future, much about data storage will change. Some of the storage technology we see today will become extinct, while others might remain.

With 4.5 billion internet users relentlessly creating new content every day, it is estimated that by 2025, the amount of data worldwide will be about 175 zettabytes.[20] Certainly, an evolution in data storage is needed. Below are some of the future data storage technologies on the horizon in the coming decades.

Helium Hard Drive

We all know that helium is lighter than air. That is why a helium-filled balloon can float. Since the architecture of any hard drive requires a very thin layer of air to act as a cushion between the spinning read/write platters, this air inside adds a considerable amount of drag when the hard drive spins at 7200 rpm; therefore, additional energy is required to spin the disk platter. Since helium is lighter than air, a helium HD can spin up to seven times faster than a conventional hard drive by replacing air with helium. This means higher storage capacity, faster processing and lower cost per gigabyte.

In 2019, WD released the world's largest hard drive using helium to power a 20 TB hard disk drive.[21] It is expected that the helium Hard Drive storage will be a trend of the immediate future.

Figure 6.12: Western Digital Ultrastar DC HC650

Source: Western Digital

[Note: The Helium HD can spin seven times faster than a conventional HD because helium is seven times lighter than air.]

DNA Digital Storage

You read it correctly. Yes. It is DNA. It exists in every cell in our body. It is the most futuristic form of data storage technology unparalleled compared to conventional storage technology. In fact, all the world's data can easily fit in a single teaspoon of DNA. DNA storage is durable. It is basically eternal.

In 2012, Harvard University researcher George Church had successfully encoded a 53,400-word book in HTML, 11 JPEG images and one JavaScript program to the DNA.[22] It works by encoding digital data into a DNA sequence in the form of A, C, G and T. The corresponding sequence information is then synthesized into an artificial DNA. To retrieve the information, the artificial DNA strand is decoded back into binary.

DNA storage is not science fiction. In fact, Google had already initiated a DNA digital data storage facility under Google Genomics.

However, DNA storage is not without its drawbacks. Writing and reading DNA is a very time-consuming process.

In March 2019, The University of Washington and Microsoft researchers described in a Nature Scientific Report paper that an automated DNA proof concept is able to encode the word "hello" in DNA, store it, and then read it back, but the write to read latency is approximately 21 hours.[23] This result shows that DNA data storage is prohibitively slow and very expensive. As I am writing now, the most cost-effective DNA storage technique costs about $3,500 per MB to write the data and $1,000 per MB to read it.

Now, the race in the future of computing storage is to build a commercially viable DNA storage technology. Catalog, one of the companies on the bleeding edge of the DNA storage technology, believes that the storage cost can be brought down to under 1/3000 cents per MB. It is developing a DNA-based data storage system to hold massive amounts of data. Its ultimate goal is to offer DNA data storage to customers that need to store petabytes of data in archives.

***

Computing memory and storage are crucial parts of computing technology. As long as technology advancement continues, they will stay important now and in the future. Despite their importance, the bottleneck in computing technology is in the processing power. After all, a faster computer is probably more important than more quickly accessible memory.

That is why I believe understanding processors, their technology and history, is a very valuable insight in the 21st century. It allows you to appreciate the technology in the future a lot more. Every single technology happening in the decades ahead will be directly or indirectly related to computer processors.

For now, we will look at the early generation of computers, how computer processors evolved, and how they will continue to reshape the digital age we live in today and the future.

Early Generation of Computers

The first generation computer uses basic components like a vacuum tube for arithmetic calculation. A vacuum tube is an electrical switch that could switch between two states: on and off. There are no physical movements in a vacuum tube. Only electrons are moving inside. It can alternate its state thousands of times faster than mechanical switches.

The Atanasoff–Berry Computer (ABC) was credited as the world's first vacuum tube computer and the first electronic digital computing device. It was built by John Vincent Atanasoff, a former Iowa State professor of physics and mathematics, and Clifford Berry, a former physics graduate student and electrical engineering undergraduate, from 1937 to 1942.[24] However, ABC wasn't a general-purpose computer like we have today. Quite the opposite, it was designed for one purpose - to break the code of the German Lorenz SZ-40 cipher machine during WWII. ABC featured about 300 vacuum tubes for control and calculation.[25] It uses binary digits to represent all numbers. All arithmetic logic functions were done using electronics instead of mechanical switches. It uses a system called regenerative capacitor memory, which was basically a pair of drums, each containing 1600 capacitors that rotated on a common shaft once per second. Each drum's capacitors were organized into 30 bands of 50 that allow the computer to do arithmetic of 30 additions or subtractions per second.[26] The primary user input was via standard IBM 80 column punch cards. The output was a front panel display.

However, ABC wasn't the only computer in the wartime period. Colossus (1943-1945) was a fully programmable electronic computer built during WWII.[27] During the war period, Adolf Hitler and his generals encoded their communication using a German stream cipher machine known as Lorenz cipher, nickname Tunny. Britain's initiative of Colossus was to invent a computer that could decode German radio messages. Like ABC, Colossus was not a general-purpose computer. It was also designed with one single purpose –code break. Colossus was designed by General Post Office (GPO) research telephone engineer Tommy Flowers to solve a problem posed by mathematician Max

Newman at Bletchley Park.[28] Its goal was to reduce the time to work out the Lorenz setting and accelerate the whole code-breaking operation; thus, shortening WWII and saving tens of thousands of lives. In fact, Colossus was the first electronic digital machine that could have a program. It is important in the timeline of digital computers.

In 1946, John Mauchly designed the world's first general purpose digital computer, ENIAC (Electronic Numerical Integrator and Computer), that forever changed the world. ENIAC had 17,467 vacuum tubes for processing, 70,000 resistors, 10,000 capacitors, 1,500 relays, 6,000 manual switches and 5 million soldered joints. It covered a 20-foot by 40-foot room, weighed 30 tons, and consumed 160kW. At that time, a computer in past generations did not have an operating system. It could only perform a single task.[29]

Figure 6.13: Marlyn Wescoff [right] and Ruth Lichterman were two of the female ENIAC programmers.

Source: spectrum.ieee.org

The Most Important Inventions in the 20th Century

By 1947, a remarkable invention happened that revolutionized the computing industry and opened the world to a digital future. In fact, it is probably the most important invention in the entire 20th century. It is the invention of the transistor – a semiconductor device that can amplify or switch electrical signals.

When vacuum tubes were used as components in a computer, they were bulky and used a lot of electrical power that ended up as heat, which shortened the life of the tube itself. On the other hand, transistors were a much more elegant solution to the needs of electronics. They are much smaller and use much less power than the vacuum tube. Because transistors use so little power, there is little heat to dissipate, and the transistor does not fail as quickly.

AT&T (Telephone and Telegraph) research and development team in the Bell Telephone Laboratory started a project seeking an alternative to vacuum tubes. John Bardeen, Walter Brattain and William Shockley were credited for the discovery of the transistor.[30] What Bardeen and Brattain had created was the "point-contact" transistor. Shockley subsequently designed a new type of transistor called the "bipolar" transistor (i.e., BJT transistor), which was superior to the point-contact type and replaced it.

The transistor works because it is a semiconducting material. In the diagram below, a small current IBE flowing from the base to the emitter opens the flow of large current ICE from the collector to the emitter. In other words, if you apply a small voltage of 0.7V between the base and the emitter, you can turn the transistor on and allow a current flow from the collector to the emitter. This is the basic mechanism of how a transistor uses small voltage for switching.

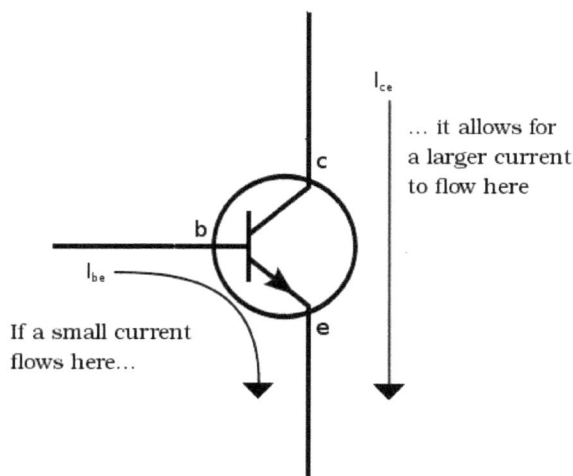

Figure 6.14: A standard NPN transistor has three pins: base, collector and emitter.

Source: Author

Transistors are made up of semiconductors such as silicon. A semiconductor is a substance that has conductivity between an insulator and a conductor due to impurity or temperature effect. The atomic structure of silicon has 4 electrons in its outer shield. Each one of these electrons is going to share with its neighboring silicon atom to form a covalent bond, which does not conduct electricity. If pure silicon has to conduct electricity, the electron must be free. So, a technique called doping is used to improve the electrical conductivity of silicon. For example, if you inject phosphorus with five valance electrons into silicon, one electron will be free. This is called N-type doping. On the other hand, if you inject boron into silicon, which only has three valance electrons, there will be a vacant position for a vacant electron to fill its position. This is called P-type doping.[31] BJTs are made of doped material and can be configured as NPN or PNP transistors. The goal is to change the way electrons behave so that we can control the conductivity to control switching.

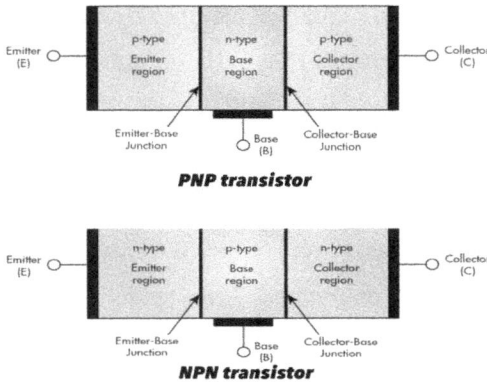

Figure 6.15: NPN and PNP transistor

Source: www.electronicdesign.com

During the 50s, there was a great technological change to move from vacuum tube to transistors. IBM 608 was the first all-transistorized calculator without any vacuum tube.[32] Because the transistor was so tiny, it enabled the world to miniaturize electronic devices.

The Rise of Silicon Valley

When we set out to create a community of technical scholars in Silicon Valley, there wasn't much here and the rest of the world looked awfully big. Now a lot of the rest of the world is here.

—Frederick Terman

When you think of Silicon Valley, you will probably think of giant technology companies like Apple, Facebook, Intel, eBay, etc.

But, do you know how Silicon Valley earned its name?

By having the highest number of innovation and manufacturers specializing in silicon-based technology. The miracle in Silicon Valley is made possible because of skilled R&D in the area, plenty of venture capital and steady US Department of Defense spending.

Stanford University was especially important in the early development of the valley. During the 1940 to 1950s, Frederick Terman, Stanford's Dean of Engineering, encouraged graduates to start their own companies.[33] He had personally invested in many of them himself. Terman is credited with nurturing Hewlett Packard, Varian Associates and many other high tech companies. He is widely credited as being the *Father of Silicon Valley*.[34]

The most remarkable achievement in Silicon Valley is perhaps the invention of transistors. It was an unprecedented development in the electronics industry, which revolutionized the 20th century. William Bradford Shockley, John Bardeen and Walter Brattain were the father or the transistor. In all began in 1947 when Shockley was put in charge of a semiconductor research group at *Bell Telephone Laboratories* in New York, belonging to A.T.&T. The purpose of the research was to solve a serious problem with long distance communication because electrical signal loses intensity over certain distance and must be amplified again. However, during that period, the amplification could only be done with vacuum valves, which consumed too much power and released a lot of heat. This drove Shockley and his team to research a more reliable amplifier in the semiconductor. After several failed attempts, the first working transistor was invented by Bardeen and Brattain was made with germanium.[35] It was called a point contact transistor and works because germanium is a semiconductor, which if properly treated, can either let a lot of current through, or let none. It was about an inch high, which is mammoth by today's standard. However, as the leader of the development team, Shockley wasn't included as the inventor. Furious, Shockley set out to design his own transistor called the bipolar or junction transistor.[36] Therefore, the transistor replaced the bulky vacuum tubes and mechanical relays and became the basic building block of modern computing technology today.

By 1956, Shockley left Bell and founded his own company called Shockley Semiconductor Labs, which was the first company to make transistors out of silicon, not germanium.

However, as a result of Shockley's personal issues and his abusive management style, eight engineers left his company and formed Fairchild semiconductor. Shockley referred to them as the "*Traitorous Eight*".[37] Among them, two of the engineers, Robot Noyce and Gordon Moore, formed Intel.[38] In 1959, Martin Atalla and Dawon Kahng invented MOSFET (Metal oxide semiconductor field effect transistor), which fueled the economic growth of Silicon Valley.[39]

The Making of the Microprocessor

Beside the invention of transistors, another important breakthrough in computing is the invention of the Integrated Circuit in the late 50s.[40]

The integration of a large number of transistors into small chips results in circuits that are much smaller, faster and less expensive than discrete electronic components.

Figure 6.16: Integrated Circuit

Source: Author

To do so, a method known as photolithography to transfer complex

patterns to a material was used. The progress of making an IC begins
with a silicon wafer with a thin layer of oxide layer added on top. A
special chemical called a photoresist will then be applied. When the
photoresist is exposed to light, a chemical reaction happens to make
it soluble so it can be washed away. A photomask is put on top of the
wafer. When applying powerful light, the mask blocks the light and
photoresist changes with the chemical reaction. A special chemical will
then wash away any remaining oxide. Afterwards, another procedure
is implemented to wash away any photoresist. A method called doping
is used. This is done by introducing impurities into the semiconductor
crystal to selectively turn different regions of the semiconductor substrate
into conductors.

Below is a patent in 1962, showing the method to manufacture a
semiconductor device, which eventually changed the world forever.

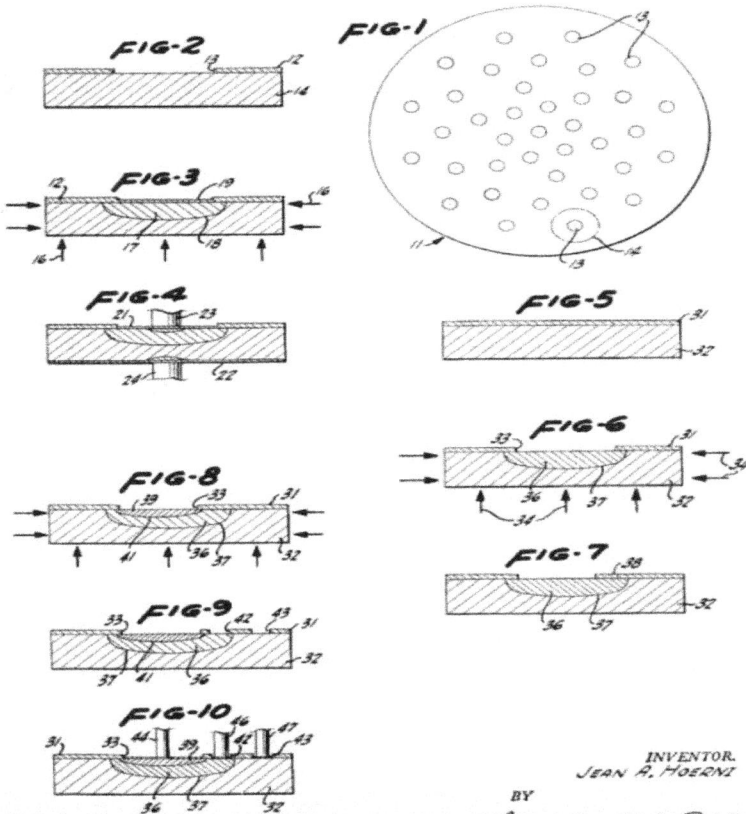

Figure 6.17: Method of manufacturing semiconductor devices

Source: Patent US3025589A

In the real world, the manufacturing process of wafers needs to be extremely precise. They cannot tolerate any particles that may fall onto the wafer and damage the chip. Technicians must wear a special uniform and undergo an air shower before being allowed inside the microprocessor factory.

The speed of the processor relies on how many transistors can be crammed into a tiny space. To get a perspective of how tiny it is, the diameter of human hair is about 80,000 nm - 100,000 nm (nanometer). A transistor today is about 5nm to 7nm. So, you can imagine even a speck of dust would affect the overall result.

[Note: A nanometer is one billionth of a meter, also expressed as 0.000000001 meters (for perspective, hair grows at roughly 1 nm per second**). In chip design, "nm" refers to the length of a transistor gate – the smaller the gate, the more processing power that can be packed into a given space.]

During the manufacturing process, a photo mask lays down millions of details at once. A single silicon wafer is generally used to create dozens of ICs. Once a whole wafer is full, you cut them up and package them into microchip.

Figure 6.18: Intel CORE i9 8th Gen

Source: Intel

As photolithography techniques improved, the size of transistors shrunk, allowing for greater density. At the start of 1960, an IC rarely contained more than 5 transistors. By the mid-1960s, an IC began to have more than 100 transistors fitted. In 1965, Gordon Moore saw the trend that the number of transistors on computer chip doubled every two years. This is well known as Moore's law.[41]

In 1968, Robert Noyce and Gordan Moore teamed up and founded a new company, combining the words Integrated and Electronic – Intel – the largest chipmaker today.

In 1971, Intel 4004, a 4-bit central processing unit (CPU) was the first processor shipped as an IC.[42] It is the first commercially available microprocessor with 2,300 transistors. By 1980, the CPU transistor count exploded to 30,000. In 1990, it increased to 1,000,000. In 2010, it is 1 billion! Computing power could be contained in a silicon chip no greater than a grain of sand.

Today, TSMC (Taiwan Semiconductor Manufacturing Corporation) is the largest dedicated semiconductor factories in the world.[43] It is the foundry of some of the best chips inside some of the best phones and computers today. In June 2020, Apple announced Mac would transition from the Intel x86 chip to making its own 5nm Silicon based ARM processor manufactured in TSMC.[44] They are transiting away from Intel because Intel is experiencing problems with its 7nm processors. In Q2 2020, Intel CEO announced that the rollout of its 7nm CPU would be delayed until late 2022 or early 2023.[45] Surely, 7nm and 5nm might not sound like a lot of difference. TSMC's 5nm technology is 15% faster and consumes 30% lower power than 7nm. But it is not just TSMC that beats Intel; AMD (Advanced Micro Devices), another major foundry in chips, also beat Intel with a CPU built on a 7nm process node – with 5nm and 3nm on the way.

So why is 7nm, 5nm, or perhaps, 3nm technology important?

As the demand for semiconductor products or services continue to grow in the 21st century, we will see an exponential rise in expectation for more powerful technological innovation and faster computers. PPA (power, performance and area) will continue to force major chip foundries for new process technology so that technology keeps pace with Moore's Law - transistor density in a device doubling every 18 to 24 months.

For this reason, understanding the future of the CPU and a bit of history of the foundry will tell us insight about where the technology trend in the future might be heading.

Inside the Brain of the Microprocessor

First, we need to understand about the architecture of microprocessor. A single silicon semiconductor chip consists of the CPU, memory and I/O (Input and Output). The system bus connects these units to facilitate the exchange of information. A system bus has a data bus, an address bus, and control bus.

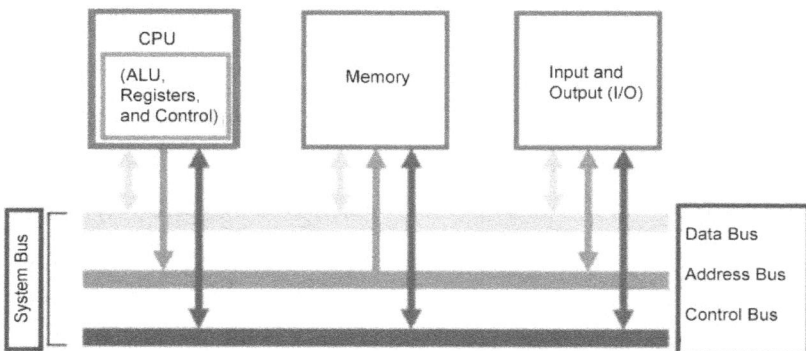

Fig 6.19: Architecture of the Microprocessor

Source: www.elprocus.com

[Note: A bus in here refers to the circuitry.]

The CPU is like the brain of the computer. It is responsible for telling other parts of the computer what to do. It is ranked in terms of hertz (e.g. 3.2Ghz), which is the cycles per second. The faster the frequency, the faster the speed, and hence, the better the performance. A CPU consists of one or more ALU (arithmetic logic unit) to perform arithmetic and logic operations. Processor registers are normally at the top of the memory hierarchy. They are measured by the number of bits they can hold. They provide the fastest way to access data. When given a specific CPU instruction, the CPU will move data between registers and memory.

Instruction	Description	Address or Register
LOAD_A	Read RAM Location into register A	4-bit RAM Address
LOAD_A	Read RAM Location into register B	4-bit RAM Address
STORE_A	Write from register A into RAM location	4-bit RAM Address
ADD	Add two registers, store value into second register	2 bit register ID, 2 bit register ID

Table 6.3

Instruction Table for Computer

Applications that run in our computer like Word or Excel aren't written in CPU instructions. They are written in high level programming languages like C++ and are compiled and decoded into micro-operation (op - code) for a specific instruction set so that they can run correctly on CPU.

[Note: micro-operation is an operation executed on data stored in registers.]

So, as you can see, from a processor's perspective, a program is merely a set of instructions executed in sequence as the CPU receives them. The CPU fetches instructions from memory, decodes these instructions and then executes those instruction. In a nutshell, all a CPU does is continuously carry out its fetch-decode-execute cycle. Inside a CPU, there can be one or multiple cores. CPU cores are the CPU's processors. More processing units allow more than one instruction to be carried out at the same time. This allows multitasking. One core might be running Microsoft Word. Another core might be running antivirus software in the background. This improves the speed and performance.

However, how efficiently and effectively it runs the instruction depends not only on the number of transistors but on the CPU architecture as well. The link between instructions and processor hardware design is what makes CPU architecture.

The Future of CPU Architecture

When the Intel 4004, a 4 bit-processing unit, was released by Intel in March 1971, it was the world's first complete general purpose CPU, with 46 instructions. It was a true milestone in computer history, and it also marked the beginning of Intel's global dominance in the processor industry. Designed in 1978, x86 architecture was one of the first instructions set architecture (ISA) for microprocessor-based computing. It defines how a processor handles and executes different instructions and passes them from the operating system and software program.

[Note: The 'x' in x86 denotes the ISA version.]

Today, there are three major makers of processors: Intel, AMD and

ARM. Each of them has designed different CPU architecture, and each of them sees the future of computing very differently.

Intel (x86) targets peak performance. Its architecture is CISC (Complex Instruction Set Computing).[46] As its name suggests, CISC has a complex instruction set. Each set of instructions has additional hundreds of separate instructions. A single instruction performs numerous low level operations like arithmetic operation, or load from memory. CISC architecture is a multi-step process. It is slow. It is hardware orientated. The average cycle per instruction (CPI) of CISC is in the range of 2 to 15.

ARM, on the other hand, uses RISC (Reduced Instruction Set Computer) architecture.[47] Unlike Intel x86, it only use simple commands that can be divided into several instructions. Each instruction is expected to perform very small jobs only. Instructions are simple and of similar length. Even for a complex command, they are broken down into small instructions and bundled together so that it can complete within one clock cycle. RISC architecture is very fast. It is software orientated. The average CPI of RISC is only 1.5.

[Note: AMD and Intel x86 CPU use CISC architecture. PIC microcontroller, ARM and AVR use RISC.]

Over the past decade, ARM has won the as most welcomed choice for smartphone devices. It uses less power and is more energy efficient. That is why Apple ditched Intel and is shifting towards building its own Apple Silicon chip using ARM-based processors.

Even so, the battle between computing architecture doesn't end here. Scientists at IBM are developing a game-changing computer architecture that can better equip handling data loads from artificial intelligence (AI). This new computer architecture is based on the human brain, particularly the memory, the neurons and synapses. Their design draws on the concept that the human brain significantly outperforms conventional computers.

Conventional computer uses von Neumann architecture that has a CPU, memory, storage and I/O device.[48] The human brain, however, does not have the same components. The researchers propose that brain-inspired computers could have coexisting memory and processing units. The pioneer, Abu Sebastian, explained that compared with supercomputers running kilowatts of power, our brain computes with merely 20 to 30 Watts only. In the brain, synapses both compute and store information. Hence, memory has a more active role to play.

Sebastian and his team of IBM researchers exploited a memory device to learn how the brain's memory and processing are collocated. Then they draw on the brain's synaptic structure as an inspiration on designing a PCM (phase change memory) device to accelerate deep neural learning.[49] (See later chapter). When they ran an unsupervised machine learning on both a conventional computer and a prototype of phase change memory device, they found that the phase change memory system could achieve 200 times faster than a conventional computing system.

[Note: Phase-change memory (PCM) is a non-volatile memory stored at nanometer scale. PCM are not only used to store data but to compute as well. It eliminates the need for transferring data back and forth between the CPU and DRAM. PCM works by exploiting the behavior of phase change material that can be switched reversibly between amorphous (high electrical resistivity) and crystalline phase (low electrical resistivity).]

***

The birth of the computer is probably the single most significant event in human history. It paves the way for all technology today and the future. This chapter has a lot of computing history about how we come to the point we are today. It also explained computer memory, storage and processors at length. In the next chapter, I am going to talk about the information age and how it will continue to reshape our life in the 21st century.

Chapter 7
A Chronicle of Information Age

Did you know that about 150 years ago, sending a mail from London to Australia would have taken two to three weeks? Today, we have instant messaging that allows us to communicate within a fraction of a second.

This is the miracle of the Internet.

The Internet brings us closer together. It has become an indispensable part of daily life.

Without the Internet, there is no modern world.

The Father of the Internet

With Moore's law, the shrinking of transistors on integrated circuits and microprocessors has created a world of explosive computer memory and processing power. The price per performance of memory (bits per dollar) rose exponentially. With the rapid expansion of the telephone network, improvements in plastic insulation, the development of coaxial cables, and advances in fiber optics, all paved the way for the birth of the Internet – a technological singularity people from the early 20th century wouldn't be able to foresee.

So, how did the Internet happen?

Until the mid-20th century, the computer was considered just a computation device for faster computation only. It wasn't until 1962 that computer scientist, J.C.R Licklider, proposed that computers could talk to one another.[1] Licklider was the director of the U.S. Department of Defense's Advanced Research Projects Agency (APRA). He envisioned the earliest ideas of global networking in a series of memos discussing an *Intergalactic Computer Network*.[2] His 1968 paper, "*The Computer as a Communication Device*," illustrates his vision of network computer application using a computer network for communication.[3] Ultimately, his vision led to ARPANET (Advanced Research Projects Agency Network) in the late 1960s. Licklider is often referred to as the father of the Internet.[4]

ARPANET

On 29th October 1969, at 10:30PM, Charley Kline, a student programmer at the University of California, Los Angeles (UCLA), sent the letter "l" and the letter "o" electronically more than 350 miles to a Stanford Research Institute computer at Menlo Park, California. UCLA became the first node of the Internet. He intended to send the word "login," but the system crashed immediately afterwards, and the Stanford computer only received the first two letters.[5]

That was when the history of the Internet was made.

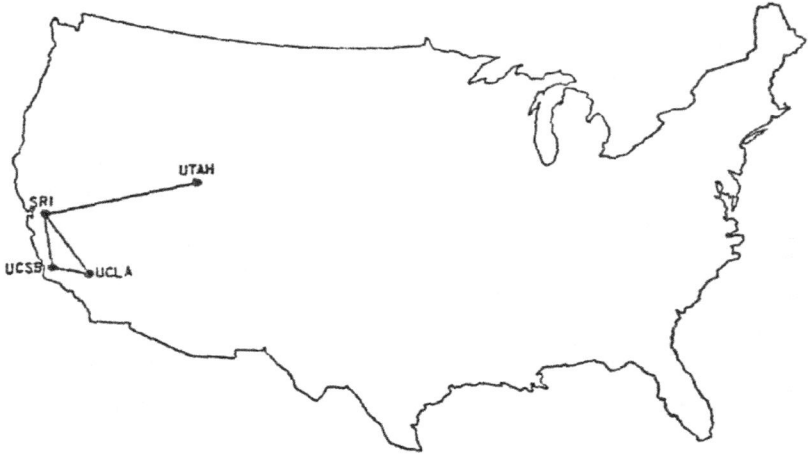

The ARPANET in December 1969

Figure 7.1: A map of the four connected computers when the first ARPANET message

was sent

Source: www.vox.com

[Note: The network only connected the University of Utah with three research centers in California.]

ARPANET was a test of a novel technology called packet switching. It breaks data into packets that can transmit efficiently.[6] One of the first ARPANET applications was Telnet.[7] It allowed computer scientists at the ARPANET site to login to a computer at another site.

By the end of 1970, ARPANET had grown to include other universities, such as MIT and Harvard. By 1973, ARPANET had become international. A satellite link was set up connecting Norway and London to the other nodes in the U.S. New applications like Email was invented in 1971 with the symbol "@" in the email address. FTP was also invented in the same year to allow ARPANET users to transfer files to each other. ARPANET allowed computer scientists who had access to the network to stay in touch. A new bulletin board system called Usenet was invented in 1980, allowing users to stay in touch and talk about different subjects, much like Facebook and Twitter today.[8]

The Birth of the Internet

As ARPANET continued to grow, the network operators soon realized that a centralized network would soon become unmanageable. They decided that ARPANET needed to decentralize. Instead of one network, ARPANET needed to be split into a network of networks. This was the birth of the Internet.

A set of network standards called TCP/IP (Transmission Control Protocol/Internet Protocol) was developed to control the format of the data packets transmitted over the network.

On 1st January 1983, the ARPANET switched to using TCP/IP and became the modern Internet we know today.[9]

A decade later, the Internet became a global network.

The Technology Behind the Internet

In its simplest form, when a computer wants to transmit data to another computer, it wires the data as an electrical signal on the cable. Since the cable is shared in a network, every computer in the network sees that data. However, they don't know whether the transmitted data is for them or for another computer. Ethernet requires each computer to have a unique Media Access Control called a MAC address. This unique address is put into the header that prefixes any data sent over the network. Computers in the network simply listen to the Ethernet cable and only process data with their address in the header.

[Note: Ethernet is the technology connecting devices in a wired local area network (LAN), enabling them to communicate. It was commercially introduced in 1980 and first standardized in 1983 as IEEE 802.3.[10]]

Today, every computer comes with a MAC address. Carrier Sense Multiple Access (CSMA) is the MAC protocol that many computers use simultaneously to sense the multiple access of the carrier. The rate the carrier can transmit is called bandwidth.

[Note: Carrier is any shared transmission medium that carries data, such as copper wire of Ethernet or the air, carrying radio waves for WI-FI.]

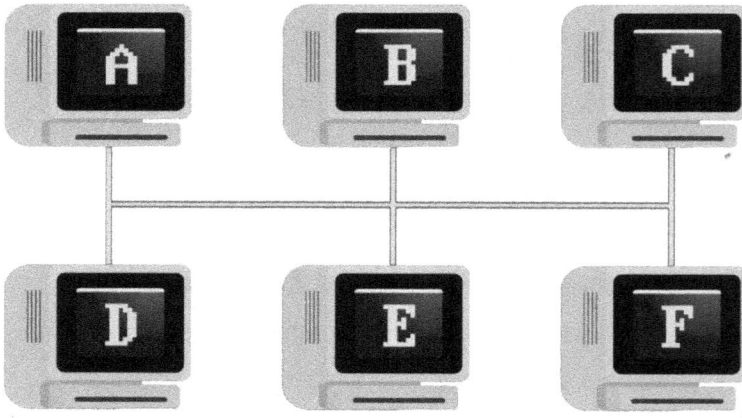

Figure 7.2: Network of computer

Source: Author

However, as the network traffic increases, the probability of two computers writing at the same time increases. This is called a collision. Soon, it will be like every computer talks to one another.

Fortunately, Ethernet has a quick fix. When a transmitting computer detects a collision, it waits for a brief period before retransmitting. This mechanism definitely helps, but it doesn't solve the problem entirely.

So, to reduce the likelihood of a collision, the network is broken down into two collision domains using a switching network. This is basically a switch that sits between two networks and only passes data if necessary. This is essentially how a big network is constructed. The Internet, the biggest network of all, literally interconnects a bunch of smaller networks that allow internetworking communication.

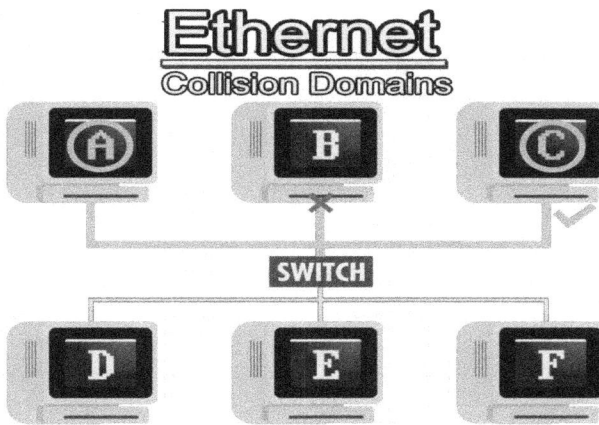

Figure 7.3: Collision and switching network

Source: Author

What is interesting about these big networks is that there is often more than one path to get data from one location to another. This leads to a fundamental concept called routing.

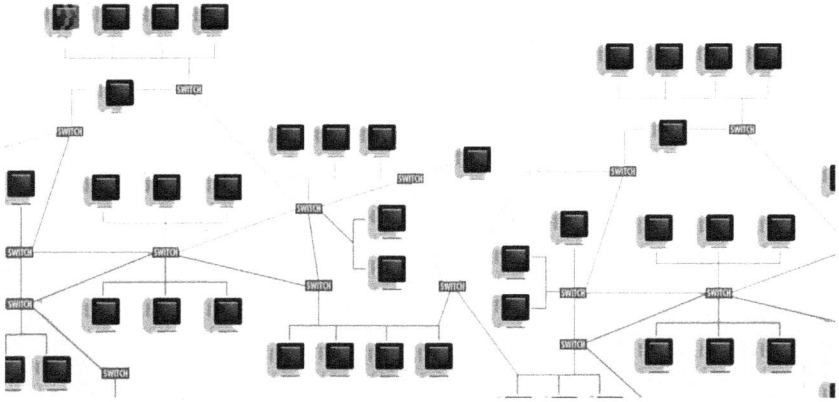

Figure 7.4: Routing

Source: Author

One routing method is called message switching.[11] Instead of a dedicated route from A to B, messages are passed through several stops. It can use different routes instead of a designated one. This makes routing more reliable and fault-tolerant. Message switching can also help to detect what is wrong with the network. Imagine cities are like message routers, and the number of hops a message takes along a route is called a hop count. Keeping track of the hop count can help to identify the routing problem. The downside to message switching is that the whole message needs to be transmitted. Because some messages might be huge, it will clog up the network and cause a traffic jam. A solution is to chop the message into many small pieces called packets. Each packet contains a destination address on the network, so the router knows where to forward them. This format is defined as Internet Protocol or IP. It is a standard created in the 1970s. Every computer connected to the network has an IP address. For example, 172.217.7.238 is an IP address for one of Google's servers.

[Note: IP address is assigned by your ISP (Internet Service Provider). It is not fixed. It can change by turning your modem or router off. When you are at a coffee shop, the IP address of your laptop changes and is assigned by the ISP for that coffee shop's internet provider.]

Network routers constantly try to balance the load across whatever routes they know. So, chopping data into small packets and passing them along flexible routes with spare capacity is more efficient and fault-tolerant. It is how the Internet runs today. This is called packet switching.

Think of the Internet as a huge distributed network that sends data around as little packets.

IP is a low protocol. A data can end up at the destination computer, but the computer might not know which application to give the data to: YouTube or Call of Duty.

IP Header	Data Payload

Hence, a more advanced protocol is needed. One of the simplest and most common is the user datagram protocol (UDP). UDP has its own header. When a packet arrives at the computer, the operating system will look under the UDP header and read the port number.

So, an IP gets the packet to the right computer. UDP gets the packet of the correct program running on that computer.

IP Header	UDP Header	Data

Before a packet is sent, the transmitting computer calculates the checksum by adding all the data together. In UDP, the checksum value is stored in 16 bits. If the receiving computer gets the packet and the checksum value is the same, it gets all the data. Otherwise, the data is corrupted. The problem with UDP is that it does not have any mechanism to fix the data or request a new copy of the data. Imagine you are receiving an email, and the middle chunk of information is missing in the middle; that will be catastrophic. That is why a transmission control protocol is absolutely necessary. People like to call it TCP/IP.

[Note: checksum is an error detecting method in networking.]

IP Header	TCP Header	Data

Like UDP, TCP has a header and a checksum. Packets are given in sequential numbers, and the receiving computer puts the packets in the correct order. TCP also requires that once a computer correctly receives a packet, which means the data passes the checksum, it sends back an acknowledgment (ACK) to the sending computer. Knowing that the packet has been sent successfully, the sending computer can send the next packet. TCP can send a lot of packets simultaneously. It can handle out of order packets, dropped packets and retransmission.

In a nutshell, when your computer wants to connect to a website, it needs an IP address and a port. (e.g., port 80 at IP 172.217.7.238). The combination of IP address and port as an endpoint is called a socket. A TCP connection is defined by two endpoints also known as sockets.

Your PC – IP1 +port 60200 ———— Google IP2 + port 80

IP1 + port 60200 = socket of the client PC

Google IP2 + port 80 = socket of the destination server

[Note: The port identifies the application using the machine. Port 80 refers to the web servers. TCP 60200 uses Transmission Control Protocol.]

In fact, if you type 172.217.7.238 in your browser, it will bring you to

Google. But it is much easier to remember google.com than a long string of digits. So, the Internet has a special service that maps the domain name to the address. It is like a phone book called the domain name system (DNS).

The Birth of the World Wide Web

The World Wide Web (WWW) is not the same as the Internet, even though people often use the term synonymously. The Internet is made up of routers and protocols. WWW runs on top of the Internet. It has millions and millions of applications running on servers all around the world, assessed by web browsers.

Before the idea of hyperlinks, every time you wanted to switch from one piece of information to another in your computer, you'd have to go through the file system or type in your search bar manually. However, this took too many steps. With hyperlinks, you can easily jump from one piece of information to another by simply clicking the link.

For webpages to be hyperlinked to one another, each hyperlink needs an address specified by the Uniform Resources Locator (URL). When you request information from a site, the first thing your computer does is look up the DNS. It takes the DNS as the input and replies with the computer's IP address. With the IP address of the computer you want, your browser opens up a TCP connection to the computer that is running the webserver. The standard port of the webserver is 80. When the connection is established, your computer connects to the webserver of the designated address. Next, your computer communicates with the webserver, requesting the hypertext page using hypertext transfer protocol (HTTP).

If a user asks for a page that doesn't exist, it returns error 404.

The very first version of Hypertext Markup Language (HTML) was created in 1990.[12]

Tim Berners-Lee wrote the first web browser and the web server in 1990. At that time, Tim Berners-Lee was working for CERN in Switzerland. He simultaneously created several fundamental web standards, such as HTTP, URL and HTML.[13]

By 1991, the World Wide Web (WWW) was born.[14]

Mosaic was a web browser created in 1993 at the University of Illinois at Urbana–Champaign.[15] It was the first web browser that allowed graphics to be embedded alongside text. It also introduced a friendly GUI interface and has bookmark functions.

By the end of the 1990s, many web browsers like Netscape Navigator, Internet Explorer and Mozilla were in use.[16] Many web servers, like Apache and Microsoft Internet Information Services (IIS) were also developed. Many new websites popped up every day. Giant websites like Amazon and eBay were founded in the mid-90s.[17]

As the web flourishes, people demand ways to find information. If you like to go online shopping and know eBay is the site for you, you can just type ebay in your browser, and the site ebay will come up. But what if you don't know the name of the site to find the information you need?

The Birth of the Search Engine

Before the search engine was developed, people hyperlinked sites to one another in the form of a directory. People maintained it manually.

But as the web grew, search engines were developed.

The first web search engine was JumpStation, created by Jonathon Fletcher in 1993 at the University of Stirling in Scotland.[18] This requires three pieces of software. The first software was a web crawler - a software that followed all the links it could find on the web. If it followed a link that had new links, it would add those to its list. The second software was an ever-enlarging index, recording when text terms appear on the page the crawler visited. The final piece of software was a search algorithm. For example, when I type dog in Jumpstation, every web page with dogs would come up in a list. The ranking of a webpage depends on the number of times a search term appears on a page. This works okay initially, but when people began to game the search engine by putting "dog" hundreds of times in their site to draw traffic, problems happen.

Google rose to fame due to its clever algorithm that sidestepped this issue. Instead of trusting a website's content, they look at how the website is linked to other sites. For example, if a spam page called www. dog.com has a lot of keyword dogs over and over again, no site will link to it. But, if a site www.dog.org is an authority on dogs, then other sites will link to it. These reputable links are called backlinks. This started as a research project called BackRub at Stanford University in 1996. Two years later, it became the Google we know today.[19]

The Infancy Information Age

The birth of the Internet is just the prelude of the transformation into the information age. Maybe you are a post-millennial and born in an era where computers and Internet are already part of your daily life. Maybe you are like me, who was born during the 80s. Maybe you are a baby boomer who witnessed the infancy state of the computer evolution. The invention of the computer and the Internet is the heart of all future technology in the 21st century.

Right now, we are twenty years into the 21st century, and we already see unimaginable changes in information technology. It has completely reshaped our lives dramatically in every way. Computing devices are no longer just limited to PC but can be other connected devices controlled by the Internet. Below are some of the exciting information age technologies that are reshaping our world in the next twenty years.

Instant Messaging Evolution

While messaging is commonplace today, it was only two decades ago when chatting with friends and strangers online was a novel concept.

In 1997, when I was still in high school, ICQ (I seek you) began to gain momentum. At that time, meeting strangers from around the world using Internet was magic to me.

ICQ was the world's first stand-alone instant messenger (IM) [20], developed by Israeli company Mirabilis in 1998. Two years after ICQ appeared, its user base had already reached 5 million users. By 2010, it grew to a user base of 41 million daily users before it declined. ICQ allows users to chat one-on-one or in groups, exchange files, and search for other users. It revolutionized how people keep in contact with one another.

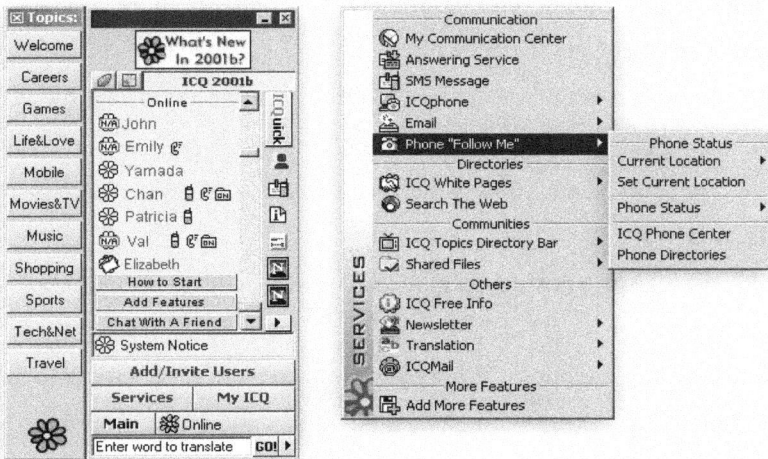

Figure 7.5: ICQ

Source: Author

IM began to revolutionize the way people communicated over the Internet. Soon, many IM applications began to pop up.

AOL launched AOL instant Messengers (AIM) in 1997.[21] One year later, Yahoo! Messenger followed. Tencent QQ was another instant messaging software like ICQ. It was first released in China in February 1999. QQ inherits functions from ICQ, it also has additional features like software skins, emoticons, chatrooms, games and Internet dating services. It grew from just 4 million users in 1999 to 823 million by 2019! [22]

To many, the 2000s were a golden era for instant messaging.

Today, over one trillion people use instant messaging. Sharing photos, making video calls and playing games became common as IM technology advanced.

Skype was launched in 2003.[23] It allowed Internet users to communicate with one another through video, voice and IM. This reduced the cost of dialing international calls. People are more connected than ever, even over long distances.

In 2006, MySpace launched the first IM platform built within a social network.[24]

In 2008, Facebook Chat allowed Facebook users to message friends in the social network.

In 2009, WhatsApp allowed users to send texts, pictures, videos and audio for free.

In 2011, WeChat, a clone of WhatsApp, was created. It became a fully integrated mobile platform with games, payments and much more.

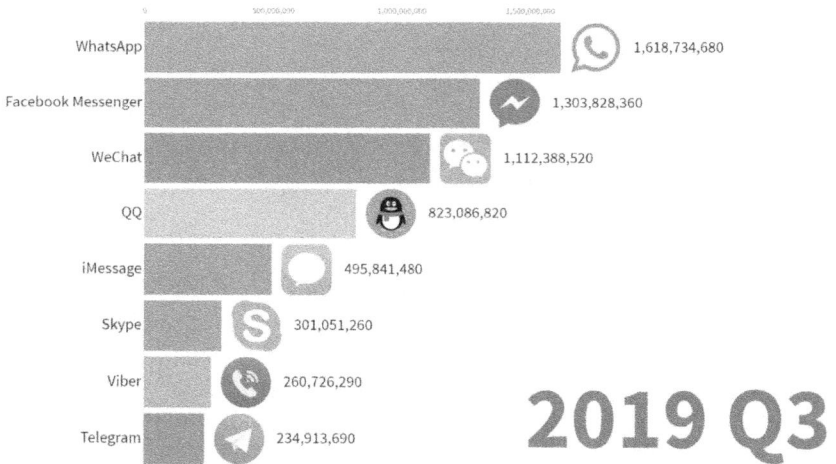

Figure 7.6: IM Statistic

Source: Youtube (Data is Beautiful)

Birth of Social Media

I was born and grew up in the 80s, an era without social media. Back then, life was very different. We met people only physically. There were no push notifications about politics. There were no stalking people on Facebook, Twitter and Reddit. The only way we get to know one another was face to face.

Figure 7.7: Social Network

Source: Author

Twenty years into the 21st century, the world changed dramatically. Social media is connecting people the way we couldn't have imagined in the 80s. If you don't know something, you no longer have to go to the library. Just Google it, and you are most likely to find pages after page of solutions posted by experts on the other side of the world for free. You can follow these people, connect with them, make friends with them.

Social Media comes in many forms. One of them is an online forum.

Early web-based forums dated back as far as 1994.[25] Online forums are like virtual communities, connecting people with the same interests. Technology, video games, sports, investing, real estate, computing issues are just a few popular areas of forum theme. People post a question on the forum, and members in the forum will answer them. Forum members will vote on the answers based on how useful they are. When combined with search engines, an online forum becomes a very powerful tool to find solutions to problems very quickly. In fact, that is exactly what I did in my first job. Instead of knowing everything to be an expert, knowing how to search and where to search for information is crucial today.

Beside online forums, the blog is one of the pillars of social media. The very first blog (Links.net), created by Justin Hall, a U.S. freelance journalist, can be traced back to 1994.[26] By 1999, the word blog had come to life.

Blogging took off in the early 2000s. Blog publishing tools like Blogger and LiveJournal began to appear, and by 2002, people started to monetize their blog using BlogAds, a precursor to Google Adsense.[27]

In 2003, Google bought Blogger and made it freely available to the world.[28] This pushed the entire concept of blogging mainstream. Anyone can publish online about anything.

In the same year, Myspace and WordPress were started. [29] [30]

Instead of just publishing, Myspace allows user to have a network of friends, personal profiles, blogs, groups, photos, music, and videos. It was the largest social networking site in the world from 2005 to 2008. I was once an active user of Myspace. I still remember how fascinating it was to publish your thoughts and photos, which could be assessed anywhere in the world, anytime, as long as I had an Internet connection. And I can change the theme or create my own theme. It was addictive.

WordPress, on the other hand, is another extremely popular blogging software. Instead of being hosted on a third party platform, you can host your own WordPress site by buying your own domain name (e.g. www. yourdomainname.com) and web hosting package. You don't need to learn web programming languages or hire a web programmer to maintain your site, as sites like Wordpress allow anyone to upload and manage any content. They call this a CMS (Content Management System). Soon, other CMS, like Joomla and Drupal, began to emerge. Although CMS has eliminated many jobs for the web designer, this has created a new type of opportunity for those who know how and where to sell themes and plugins for these CMS.

Another famous social media is YouTube. Started by former Paypal employee Jawed Karim, Steve Chen, and Chad Hurley, YouTube first launched in 2005 and has now become one of the most visited websites in the history of the Internet.[31] Google acquired it in 2006 for around $1.65 billion.[32] Unlike IM, forums and blogs, YouTube allows sharing of video with others over the Internet about anything and monetize it by attracting views.

Facebook is the most popular social networking site, founded in 2004 by Mark Zuckerberg, Eduardo Saverin, Dustin Moskovitz, and Chris Hughes, all of whom were students at Harvard University.[33] Facebook is the largest social networking website in the world. People and businesses are able to connect with one another in a way unimaginable before.

After the success of Facebook, many social networking sites appeared. Each of them targets a niche. Twitter, for example, is a micro-blogging website that acts like the SMS of the Internet. It was founded in 2006. Twitter Users can send short messages, only up to 140 characters, called tweets.[34] Reddit, founded in 2005, is another social networking site focusing on social news and media news aggregation.[35] Pinterest, founded in 2009, allows people to discover information on the WWW using images, GIFS and videos.[36] Instagram, founded in 2010, is a photo-video sharing social networking site owned by Facebook.[37]

With smartphones, people can social network with one at any time as long as there is Internet access. In just a span of 20 years, from personal to business, social media has completely transformed our lives in every way. This is happening at an accelerating rate.

Evolution of eCommerce and online shopping

Another major way the Internet transforms our daily life is the way we shop. In the past, we'd go to brick and mortar shops to buy things. Today, we can sit comfortably on our sofa in our pajamas, browsing through sites like Amazon and eBay, and with the click of the buy button, our goods will turn up within a few days. We can shop anywhere, anytime, and transact with confidence with any seller in the world.

It may surprise you that the concept of online shopping dates back fifty years ago. The first major eCommerce company was CompuServe, founded in 1969.[38] In fact, CompuServe was the first major commercial online service provider in the U.S., apart from Prodigy and America Online. This was before the widespread adoption of the Internet and WWW.

One year later, Michael Aldrick invented electronic shopping, later known as eCommerce.

In 1992, Charles M. Stack launched Book Stacks Unlimited as an online bookstore in 1992, three years before Jeff Bezos introduced Amazon.[39]

But eCommerce really didn't take off until the mid-90s.

In 1995, Pierre Omidyar introduced AuctionWeb, which later became the eBay we know today.[40] In the same year, Jeff Bezos introduced Amazon as an eCommerce platform for books.[41]

In 1998, Paypal launched the eCommerce payment system.[42]

In 1999, Alibaba was launched as an online marketplace, which turned into a major B2B, C2C and B2C platform.[43]

If you look at the annual revenue of major eCommerce platforms below, you will realize that eCommerce started off slowly in 1996, and it wasn't around 2010 before it began to pick up pace.

Annual revenue in billions of U.S. dollars

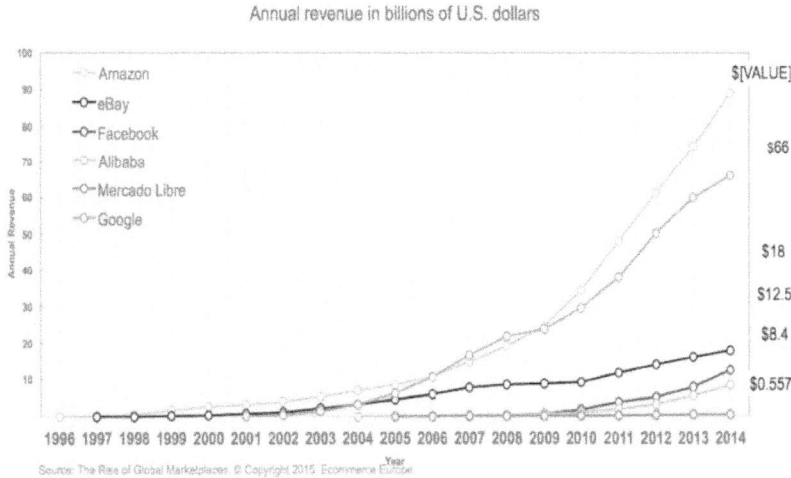

Figure 7.8: Annual Revenue of web companies

Source: The Rise of Global Marketplaces © 2015

Just last year, Amazon eclipsed Microsoft to become the world's most valuable listed company in the world.[44]

Amazon's rise to the top

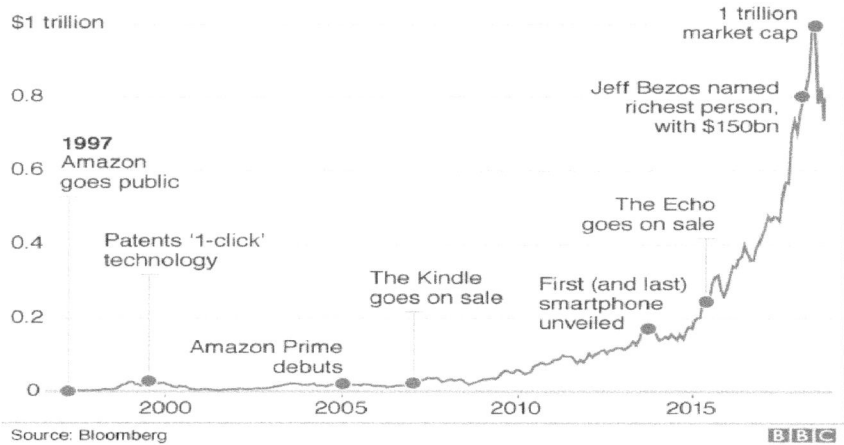

Figure 7.9: Amazon's rise to the top

Source: Bloomberg

Because eCommerce became such a big thing, everyone wanted to jump onboard. However, not many people have the technical expertise and know how to code an eCommerce website, and eCommerce technology platform providers appeared to fill this gap. They provide a 100% bootstrapped eCommerce storefront platform so anyone can own their own online store. Shopify (2006), Magento (2008) and BigCommerce (2009) are just a few of them.

Besides Paypal, Google Wallet (2011) was rolled out as a peer to peer service to allow people to send and receive money from mobile or desktop to either sender or receiver. And Facebook advertising, if used correctly, can help individuals to do the heavy lifting work of reaching a target market.

Even so, eCommerce didn't end here.

An old business model called dropshipping changed the way we do eCommerce dramatically.

In 2010, Alibaba released AliExpress.[45] On the website, individuals could purchase items from Chinese manufacturers very inexpensively and dropship the same item by listing it on eBay or their own eCommerce store on Shopify at a higher price. Dropshippers do not need to stock any products; they do not need to worry about the shipment and returns, as the manufacturers take care of this. All they need is to do market research to work out a profitable market niche that is still in Blue Ocean and advertise using Facebook ads correctly.

Evolution of the Gig Economy

In the mid and late-90s, the world saw a boom in the number of freelancers actively working. Back then, people saw freelancing as a way to make pocket money, and a gig is used to define that discrete freelance job. It can be anything from translation, proofreading, marketing, graphic design, making videos, programming and consulting, etc.

Founded in 1998, Elance was one of the early freelance sites online.[46] It was a global freelancing platform, connecting business and independent professionals remotely.

oDesk, another freelancing site, was founded in 2003 by Odysseas Tsatalos and Stratis Karamanlakis.[47] They wanted to work together, but one of them was in the U.S. and the other was in Greece. They founded oDesk to solve this problem, and it became a marketplace to allow anyone to search and hire anyone online.

The term "gig economy" didn't really take off until around the height of 2008-2009.

Then we have freelancer.com, founded in 2009, a marketplace where employers and employees can find one another. And then Fiverr, an online marketplace for a $5 freelance service, was founded in 2010.[48]

Today, Upwork, (formerly Elance and ODesk), have become the world's largest freelance marketplace.

These freelancing sites are changing the world of work in the information age. With a laptop and Internet connection, anyone can work and earn money anytime, anywhere. Unlike traditional employment, freelancers get to live the lifestyle they want. It is giving freelancers the flexibility of work so that they can spend more time with family. I was one of them.

So, do you see how the Internet is redefining work?

It is just the beginning.

An online marketplace is only one of the evolutions in the information age. The real technology behind it is P2P (Peer to Peer) network.

Evolution of Peer to Peer (P2P)

When you browse a website, your open your browser, which sends a request for data to the server through the Internet. In this case, the website works as a server, and your computer works as a client. The server accepts the request, processes it, and sends the data packets requested back to the client. The server can manage numerous clients at the same time. This is a client-sever relationship with a centralized communication model.

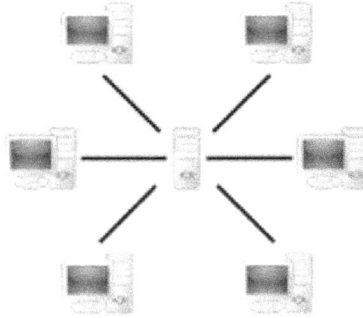

Figure 7.10: Centralized Network

Source: Author

When you download the same file from a P2P network, the download is performed very differently. Instead of downloading a file directly from the server, the file is downloaded to your computer in bits and parts that come from many other computers that are also connected to the same P2P network. At the same time, the file is also sent (uploaded) from your computer to other devices that are asking for it. You can think of a Peer to Peer (P2P) network as a distributed application architecture that partitions tasks between peers. The privilege of each peer is the same. There is no primary administrator device in the center of the network. It is decentralized.

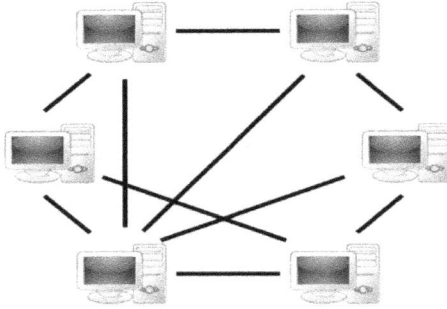

Figure 7.11: P2P Network

Source: Author

The precursor of the P2P network appears to be USENET, which was developed in 1979.[49] Basically, this is a system that allowed users to read and post messages or news. It works pretty much like online forums today except that it did not rely on an administrator. USENET copies the same news to all its servers found in the network. It works just like a P2P network distributes and uses all the resources available.

Napster was the next big thing that went mainstream in P2P history. Founded as a pioneer in P2P file sharing, Napster was used by people in music distribution and download.[50] However, the music shared on Napster was usually copyrighted and illegal to distribute. So, authorities ultimately shut down Napster because all the content shared was illegal.

BitTorrent is a famous example that demonstrates how P2P works.[51] BitTorrent is a P2P protocol. A group of computers download and upload the same torrent, transferring data between each other without the need of a central server. Here is how it works.

When a computer joins a BitTorrent swarm, it has to load a .torrent file into a BitTorrent client – a piece of software that accepts a torrent file and begin downloading data associated with it. Think of a BitTorrent client as a web browser which you need to browse the web. A BitTorrent client is needed to make sense of a BitTorrent file. Next, the BitTorrent client contacts a "tracker" specified in the .torrent file. A tracker, which is a special server, will then keep track of the connected computers. It shares those computers' IP addresses with other BitTorrent clients in the swarm, allowing them to connect to each other. Once connected, a BitTorrent client downloads bits of the files in the torrent in small pieces, downloading all the data it can get. Once the BitTorrent client has some data, it can then begin to upload that data to other BitTorrent clients in the swarm. In this way, everyone downloading a torrent is also uploading the same torrent. This speeds up everyone's download speed. If 10,000 people are downloading the same file, it doesn't put a lot of stress on a central server. Instead, each downloader contributes upload bandwidth to other downloaders, ensuring the torrent stays fast.

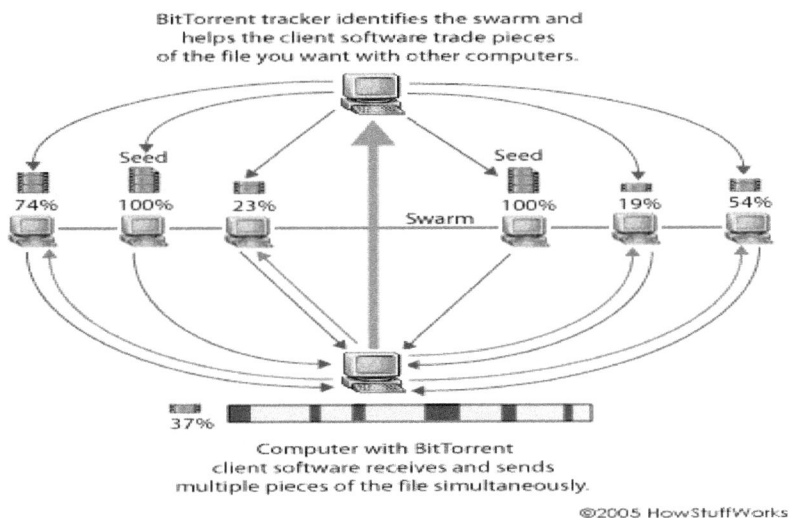

BitTorrent tracker identifies the swarm and helps the client software trade pieces of the file you want with other computers.

Seed   Seed

74%   100%   23%   Swarm   100%   19%   54%

37%

Computer with BitTorrent client software receives and sends multiple pieces of the file simultaneously.

©2005 HowStuffWorks

Figure 7.12: BitTorrent

Source: umkc.edu

P2P movement allows millions of Internet users to directly connect and collaborate with one another. Today, P2P services have moved beyond purely Internet based. It has evolved into an economic model that provides sharing access to goods and services often facilitated by a community-based online platform. It connects buyers and sellers. It is called the sharing economy.

Uber is a multinational ride hailing company, founded by Garrett Camp and Travis Kalanick in 2009.[52] Uber doesn't own any vehicles. It is simply a smartphone app that connects driver-partner and riders. As of 2019, Uber is estimated to have over 110 million worldwide users. It became the highest valued private startup company in the world.

Bitcoin is a P2P digital form of money invented by Satoshi Nakamoto.[53] It uses P2P technology to operate with no central authority or banks to manage transactions. Bitcoin can be transferred from one person to another through a P2P network. It uses a distributed ledger called blockchain - a P2P architecture that allows Bitcoin to be transferred worldwide, without intermediaries or any central server. Anyone can set up a Bitcoin node if they wish to participate in the process of verifying and validating blocks.

From file sharing to ridesharing and now to money, P2P technology is revolutionizing the world in many ways. It reduces the cost by removing intermediates. Just like the Internet, P2P connects people together through collaboration between computers. Today, many people predict a great future for the P2P network. New P2P technology will continue to affect our life in every way.

The Evolution of Internet of Things (IoT)

The Internet of things (IoT) is influencing our lifestyle in every way. From air conditioners that you can control with your smartphone to smart cars providing the shortest route or your smartwatch tracking your daily activity, IoT is a giant network with connected devices. These devices gather and share data on how they are used and the environment in which they are operated. It is all done using sensors embedded in every physical device. It can be your mobile phone, electrical appliances, traffic lights and almost everything you can do in day to day life. These sensors emit data continuously about the working state of the device.

IoT provides a common platform for sharing this huge amount of data from devices and putting this data to our benefit. It has a common language for all devices to communicate with each other. Data is sent from different devices and sent to the IoT platform. This data provides analytics on the data and extracts valuable information as per requirement. The final result is shared with other devices for better user experience, automation and improved efficiency.

For example, in an air conditioning manufacturing industry, the manufacturing machines and belts have sensors attached. They continuously send data regarding the machine's health and allow the manufacturer to diagnose issues beforehand. A barcode is attached to each product before leaving the belt, containing the product code and manufacturing details, etc. The manufacturer uses the data to identify where the product was distributed and track the retailer inventory. So, the manufacturer can make the product run without running out of stock. The air conditioner's compressor had an embedded sensor that emits data regarding its health and temperature. This data is analyzed continuously, allowing customer care to contact you for the repair work in time. This is just one of a million scenarios of how IoT works. IoT is transforming our lifestyle and how we interact with technology.

Raspberry Pi is a great platform for building IoT.[54] It is a single microcomputer with onboard wifi for direct connectivity to the Internet to send and receive data. Multiple sensors can be connected to it through GPIO (General Purpose I/O).

Figure 7.13: Raspberry Pi
Source: www.raspberrypi.org

VR and AR

From panoramic paintings to stereoscopic photographs, we have been trying to further our attempts to immerse ourselves into fictional worlds. And in the 20th century, we have numerous attempts to use film and television to bring ourselves further in.

VR (Virtual reality) is the newest technology that allows us to immerse ourselves in an interactive computer generated environment. Virtual Reality is an artificial, computer-generated world that people can experience and interact. Users can use special equipment like a VR HMD (Head Mounted Display) to completely immerse themselves in this virtual world. This contrasts with AR (Augmented Reality), where users experience the real world, but with a layer over it. Microsoft Hololens and Pokemon Go are examples of AR.

But VR is not only limited to games. VR can give you the perspective of what it is like to be in a specific place. You can potentially place a camera in the middle of Tokyo or other places in the world to give people a sense of what it feels like to be living in other places. It can also be used as a historical reenactment, so future generations can experience historical moments like the Moon Landing or the first Continental Congress to actually feel what happened. Museums, media, virtual shopping centers, training, teaching, will be a new thing in the future when integrated with VR technology.

5G Technology

Every new generation of wireless network delivers a faster speed and wireless function. 1G brought us the every fist cellphone, 2G allowed us to text for the first time, 3G brought us online, 4G brought us the speed we enjoy today. But as more and more people come online, the 4G network is just about to reach its limits.

5G will be able to handle one thousand times more than today's network, and it will be ten times faster than 4G LTE. The average speed over a 4G network is about 10Mbit per second. 5G is one Gigabit per second, one hundred times faster in connectivity. 5G only has a millisecond delay. It is like real-time. Surgeons can perform with a surgeon on one side of the world and the patient on the other. To put it into perspective, you can download a HD movie in under a second. 5G will be the foundation for virtual reality, autonomous driving, IoT, and many more future technologies in the 21st century.

But what exactly is 5G?

Our smartphone at home uses very specific frequencies on the radio frequency spectrum. (i.e., 3kHz to 6GHz). However, as more and more devices crowd the same spectrum, we are going to see slower service and more dropped connections. So, the solution is to open up some new real estate in the frequency band. Researchers are experimenting with broadcasting shorter millimeter waves, those falling between 30GHz and 300Ghz. This section of the spectrum has never been used before for mobile devices. So opening this new spectrum means more bandwidth for everyone. But millimeter waves can't travel well through buildings and other obstacles, and they tend to be absorbed by plants and rain. To get around this problem, we need a small cell network.

Today's wireless network relies on large, high-powered cell towers to broadcast their signal over long distances. But the higher frequency has a harder time travelling through obstacles. A small cell network solves that problem by using thousands of low powered mini based stations. These base stations will be much closer than traditional towers, helping it to transmit signals around obstacles. This is especially useful in the city. When a user moves behind an obstacle, his smartphone will switch between base stations in better range of his device.

MIMO (multiple-input multiple-output) is another pillar to 5G technology.[55] 4G base stations have dozens of ports for antennas that handle all cellular traffic, but a massive MIMO base station can support a hundred ports. This can increase the capacity of today's network by a factor of 22.[56] Massive MIMO is a key enabler of 5G's extremely fast data rate. It increases the network capacity, improves coverage and user experience.

Today's antenna broadcasts information in every direction at once, and all of those crossing signals will cause serious interference. This is why 5G requires beamforming. Beamforming is like a traffic signaling system for cellular signals.[57] So, instead of broadcasting to a wide area in every direction, it allows a base station to send a focused stream of data to a specific user. This precision prevents interference and is way more efficient. A station can handle more incoming and outgoing data at once.

***

The exponential growth of the Internet is transforming everyone's daily lives. It evolved from ARPANET in the 1960s into a worldwide grid of knowledge and resources that connect everyone in the world today. Alongside an exponential growth in telecommunication and broadband technology, it brings the world together. The growth of price-performance of Internet-related technologies has doubled every year, enabling global commerce at an accelerating rate. Not only that, the advent of this global, decentralized network is a pervasive, unparalleled democratic force. It is a movement towards more capitalism, economic growth and prosperity.

As Internet traffic and resources grow, the web is becoming a global knowledge base. We are entering a new paradigm where machines can read, analyze and understand data from this knowledge base to make meaningful decisions. The core technology of the computer and the Internet marks the beginning of the rise of artificial intelligence.

Chapter 8
Rise of Artificial Intelligence

"In order to change an existing paradigm you do not struggle to try and change the problematic model you create a new model and make the old one obsolete"

-Buckminster Fuller

In 1956, an American computer scientist, John McCarthy, coined the term "artificial intelligence" and organized the famous Dartmouth Conference.[1] He proposed a system that could make machines reason like humans, capable of solving problems, abstract thoughts and self-improvement.

Although John McCarthy is renowned as the father of artificial intelligence, the journey to understand if a machine can think began way before that.

The Turning Test

Imagine if you are inside a bare room. In the center of the room are a table with paper, pencil and a Question and Answer book in Chinese characters. Now, imagine if there is someone outside the room who could speak Chinese, but she can only communicate with you by exchanging papers in Chinese characters through a slot in the door. Your task is to look through the Question and Answer book in Chinese characters and reply in Chinese.

At the end of the experiment, the Chinese lady will think you speak perfect Chinese. She will think you understand the language perfectly.

But were you thinking during the communication? Do you understand what the questions were?

The truth is, you don't speak a word of Chinese. The Chinese characters were nothing more than alien symbols to you.

All you did was look up the Question and Answer book in Chinese.

So, if a computer simply follows instructions, it isn't really thinking.

But what is thinking? Can artificial intelligence really think?

British scientist Alan Turing decided to disregard all these questions in favor of a much simpler one – can a computer talk like a human?

This question led to measuring artificial intelligence that has famously known as the Turing Test.

In the 1950 paper, "*Computing Machinery and Intelligence*," Turing proposed the following game.[2]

A human judge has a text conversation with two unseen players and evaluates their response. One of the players is a computer, and the other is human. To pass the test, a computer will have to convince the judge that it is a human player. In other words, a computer would be considered intelligent if its conversation couldn't be easily distinguished from a human's.

Turing has predicted that machines with 100MB of memory would be able to easily pass the test.[3] However, even today's computer has far more memory than that, and few have succeeded. Those who have done well focused on finding clever ways to fool the judge rather than using computational power.

The first program that claimed to have passed the Turing test was ELIZA.[4] It managed to mislead people to talk more by mimicking a psychologist. Another program called PARRY succeeded in the test by steering the conversation back to its preprogrammed obsessions.[5]

The computers succeeded in fooling the judge, which highlights one weakness of the test – humans always attribute intelligence as a whole range of things that aren't actually intelligence.

After that, competition like Loebner Price made the test more formal, where the judge knows some of their conversation partners are machines ahead of time. Many chatbot programmers have used the same strategies to ELIZA and PARRY. In 1997, winner Catherine could carry on an amazing, intelligent conversation. The recent winner, Eugene Goostman, was given the persona of a 13-year-old Ukrainian boy.[6]

Meanwhile, other programs like cleverbots use a different approach. They statistically analyze a huge database of real conversation to determine the best response. Some of them stored previously recorded conversations and improved them over time. While a cleverbot's response sounds incredibly human, its lack of personality and inability to deal with a brand new topic is a dead barrier.

It looks like carrying a natural human conversation requires more than just increasing processing power and memory.

Machine Learning

"I think people need to understand that deep learning is making a lot of things, behind-the-scene much better."

Geoffrey Hinton - Godfather of deep learning.

Computers are incredible at storing, organizing, fetching and processing huge volumes of data. That is perfect for website like Amazon with millions of items for sale, but what if we want to use a computer not just to fetch data but to actually make decision based on the data? This is the essence of machine learning – algorithms that give the computer the ability to learn from data and then make predictions and decisions.

Programs with this ability to make decisions are extremely useful in answering questions like: Is this a spam email? Is this person suffering from heart disease? What YouTube video do you recommend?

While these suggestions are useful, we can't call these programs intelligent in the same way as human intelligence. So, even though we use the term artificial intelligence, most computer scientists would agree that machine learning is a set of techniques for AI.

Machine learning and AI algorithms seem to be sophisticated. So, to understand this, let's look at this simple example to see how it actually works.

Suppose Mike loves listening to music. He likes or dislikes songs based on the song's tempo and intensity. Suppose Paul loves fast and soaring intensity songs and dislikes relaxed and light intensity songs? Then the chart below can describe Mike's preference.

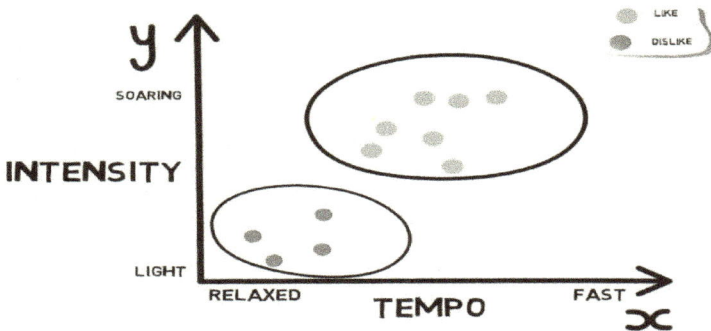

Figure 8.1: Machine Learning

Source: Author

Suppose song A has a fast tempo and soaring intensity? Can you guess whether Mike likes it or not?

You are correct. Mike likes it.

By looking at Mike's past choices, we are able to guess if Mike will like the song or not very easily.

Now, suppose Mike listens to a new song call song B, which has medium tempo and intensity. Can you guess whether Mike will like this new song or not?

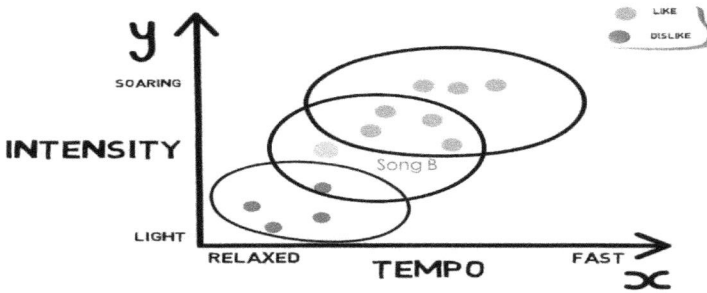

Figure 8.2: Machine Learning

Source: Author

That is where machine learning comes into play.

If you go for the number of votes encircled in song B, you can safely say Paul will like this song. It is called k-nearest neighbor algorithm.

So, instead of a computer executing a program to return a result, they predict it based on data given. More data will give a better model and a higher level of accuracy.

But what if other criteria like genre are also taken into consideration? The 2D lines will become a 3D plane, creating a decision boundary in three dimensions. These boundaries might not be just straight, either. And with more and more criteria added, you would agree that it is going to get very complicated, and that is what many real-world machine learning problems face.

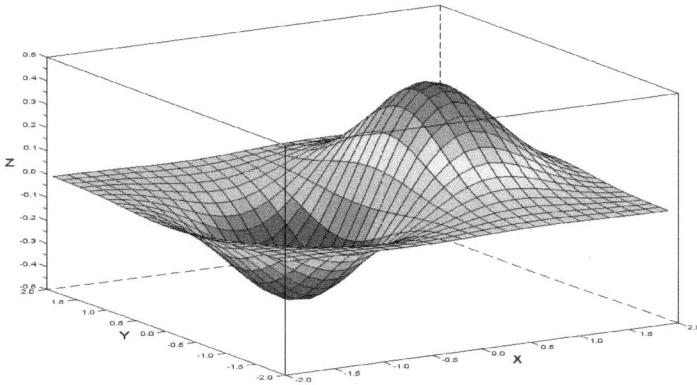

Figure 8.3: 3D plot

Source: Author

Computers with some clever algorithms can do this analysis all day long in companies like Google, Facebook and Amazon.

One of the most notable techniques to do this is artificial neural intelligence, inspired by neurons in our brains! Neurons are cells that transmit messages using electrical and chemical signals. They take signals from one cell, process the signal, and emit their own signal. These form huge interconnected networks that can process complex information.

Artificial Neurons work in a similar fashion.

Each artificial neuron takes a series of inputs, combines them, and emits a signal.

Biological Neuron versus Artificial Neural Network

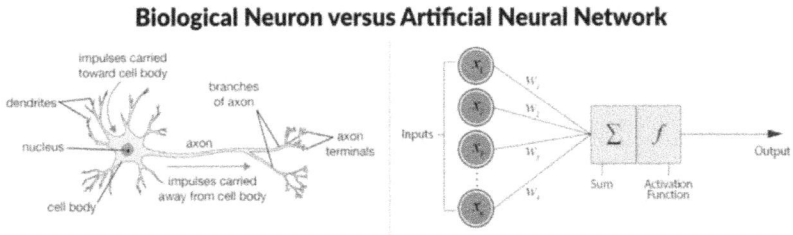

Figure 8.4: Biological Neuron vs Artificial Neural Network

Source: Author

But, rather than being electrical and chemical signals, artificial neurons take numbers in and split numbers out. They are organized into layers connected by links, forming a network of neurons.

So, if you apply this method to our example, our first layer, the input layer, provides data from Mike's preference in music. Again, we use intensity and tempo. In the end, we have an output layer that tells us whether Mike likes this music. In between, we have a hidden layer that transforms our input to output and does the hard work.

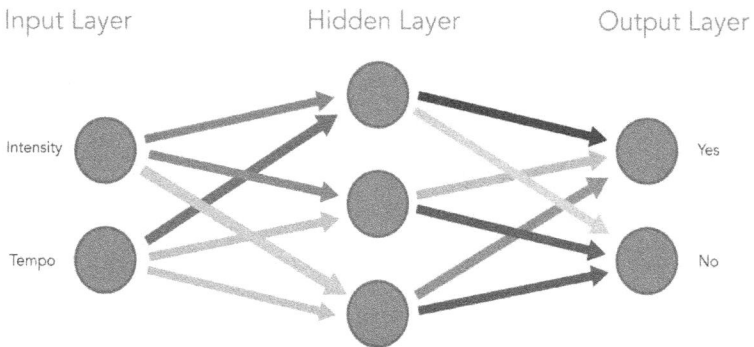

Figure 8.5: Neuron Network

Source: Author

To see what is working behind the scenes, let's zoom inside one of our neurons. The first thing that a neuron does is multiply its input by a specific weight. Let's say, 2.8 for input 1 and 0.1 for input 2. Then, the sum of its weight together is 8.4. The neuron will then apply a bias, say -3.4, to this figure for adjustment for a new value of 5, which is the applied activation function.

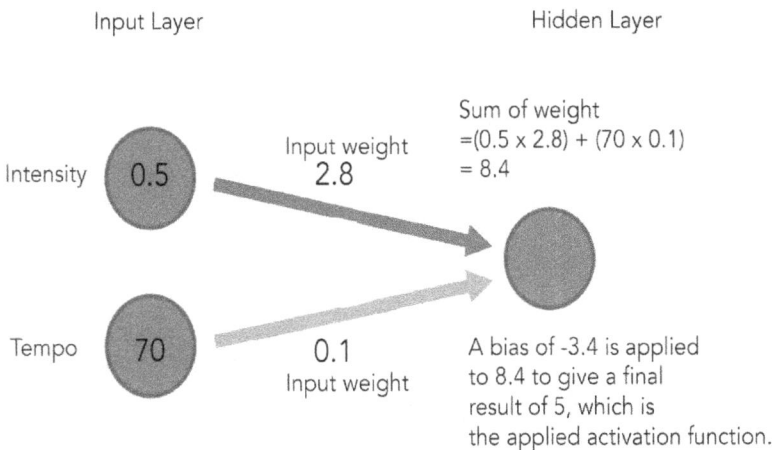

Input Layer

Hidden Layer

Intensity   0.5

Input weight
2.8

Sum of weight
$=(0.5 \times 2.8) + (70 \times 0.1)$
$= 8.4$

Tempo   70

0.1
Input weight

A bias of -3.4 is applied
to 8.4 to give a final
result of 5, which is
the applied activation function.

Figure 8.6: Neuron Training of one neuron.
Source: Author

These input weights and biases are initially set to random values when a neural network is created. The algorithm goes in and starts tweaking those values to train the neural network. This process happens over many interactions and gradually improves accuracy. This is much like how a human learns.

So, the process of summing, weighting, applying bias and applied activation function is computed for all neurons in one layer, and the value propagates forward in the network, one layer at a time.

In reality, the hidden layer doesn't have to be just one layer. It can be many layers deep. This is where the term deep learning comes from.

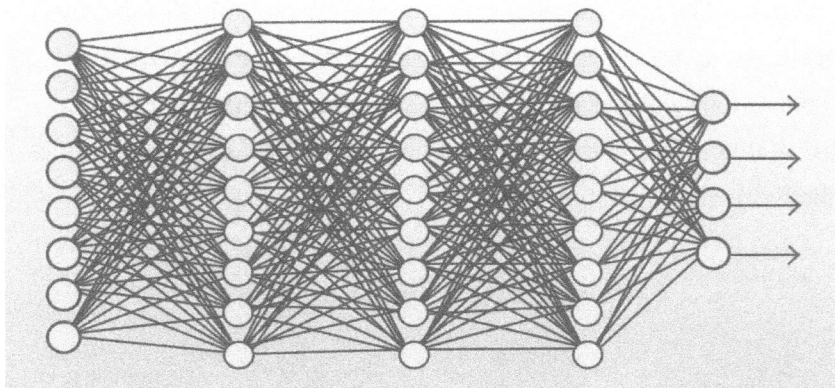

Figure 8.7: Deep Learning

Source: Author

A couple of years ago, Google and Facebook demonstrated a deep neural network that could find faces in photos, as well as humans, with high accuracy. [7] [8] AlphaGo and Deepblue use similar AI learning techniques to beat world champions in the same way.[9] From predicting and identifying disease to revolutionizing the way we work, deep neural networks are used in self-driving cars, healthcare, entertainment, automatic handwriting generation, and many other industries.

When we think of artificial intelligence, we often look at robots replacing repetitive jobs with manual labor or data-optimization. However, it might surprise you that a new wave of emerging AI is about to change your perspective.

Aiva Technologies is one of the leading startups in AI music composition. They have created an AI called "Aiva" (Artificial Intelligence Virtual Artist) and taught it how to compose classical music.[10] Aiva's musical piece wasn't just an elementary level. In fact, the music Aiva composed gained the worldwide status of composer. It was registered under the France and Luxembourg authors' right society (SACEM), where all of its works reside with copyright to its own name. Some of its work was even used in games studios and films.[11]

Aiva achieved this remarkable result by deep learning algorithms that use reinforcement learning techniques. This allows Aiva to understand and model high-level abstraction in data, such as pattern in melody. After listening to a large amount of music and learning the models of music theory through training, Aiva is able to compose a classical melody in a matter of minutes.

Deepfake is synthetic media in which a person in an existing image or video is replaced with someone else.[12] Deepfake gains its name by using deep learning techniques to fake. It relies on a type of neural network called autoencoder. It uses an encoder that encodes an image by reducing it to a low-dimensional latent space and a decoder to reconstruct the image from the latent representation. Deepfake is widely used in pornography and politics, mainly for fraud and malicious use or to boost the popularity of the video.

Figure 8.8: Deepfake
Source: Youtube (TheFakening)

Computer Vision

Computer vision (CV) is one of the critical components of artificial intelligence. Computer scientists have been trying hard to give computer CV for half a century. The goal of CV is to give the computer the ability to see the world analyze visual data, and then make decisions from it or gain an understanding of its environment.

Early experiments in CV happened back in the 1950s.[13] It uses a neural network to detect an object's edge to sort simple objects like circles and squares. As the Internet became popular in the 1990s, making a large set of data available for analysis and facial recognition software began to flourish. This growing set of data allows machines to identify specific people in photos and videos. In less than a decade, the accuracy rate of object identification went from 50% to 99%. Apps like Google Lens can be used to scan and translate text, identify objects, look up books, music, movies just by scanning it.

The computer's ability to see and make decisions is crucial for future AI technology. Take self-driving cars as an example. Detecting objects, landmarks, traffic signs are necessary to drive safely. Real-time sport tracking using CV is a game-changer, as a player's performance and rating will be available in real-time. In agriculture, an autonomous harvester will be able to analyze grain quality during harvest, as well as find the optimal route. For manufacturing, CV can observe and make predictive maintenance to interfere before a real breakdown happens. These are just a few examples of what can be done with CV.

But how does CV works?

Images on computers are stored in a grid of pixels. Each pixel is defined by a color, stored by a combination of three additive primary colors (RED, GREEN, BLUE) – the RGB. For example, for the white color, R=255, G=255, B=255; black has R=0, G=0, B=0. By combining different intensities of these three colors, we can represent any color.

Perhaps the simplest CV algorithm is to track a colored ball. The first thing we do is to record the ball's color, the RGB value of the centermost pixel. With this RGB value saved, we can give the computer program an image and ask it to find the closest color match. The computer will then search from top left to bottom right to check each pixel across the screen, one at a time. They will calculate the difference in RGB color and find the closest match. This color-tracking example can search pixel by pixel because colors are stored inside each single pixel. However, this approach will not work if we are trying to track something larger than a single color pixel, like the edge of an object, making up many different pixels. So, to identify such an object, the algorithm needs to look at patches, made up of many pixels.

Say, for example, we want to zoom in at the picture and look at the edge of the middle pillar.

Figure 8.9: Pillar

Source: Author

We can easily see where the edge of the pole starts. We can set up rules to determine if a pixel is an edge by asking question like – what is the difference in color between some pixels to its left and right? If the difference is small, it is probably not an edge.

We label and mark each pixel with a grayscale value. Next, we center our kernel to our point of interest, say the yellow box with value 224. After doing the math, the initial value 224, will become 147.

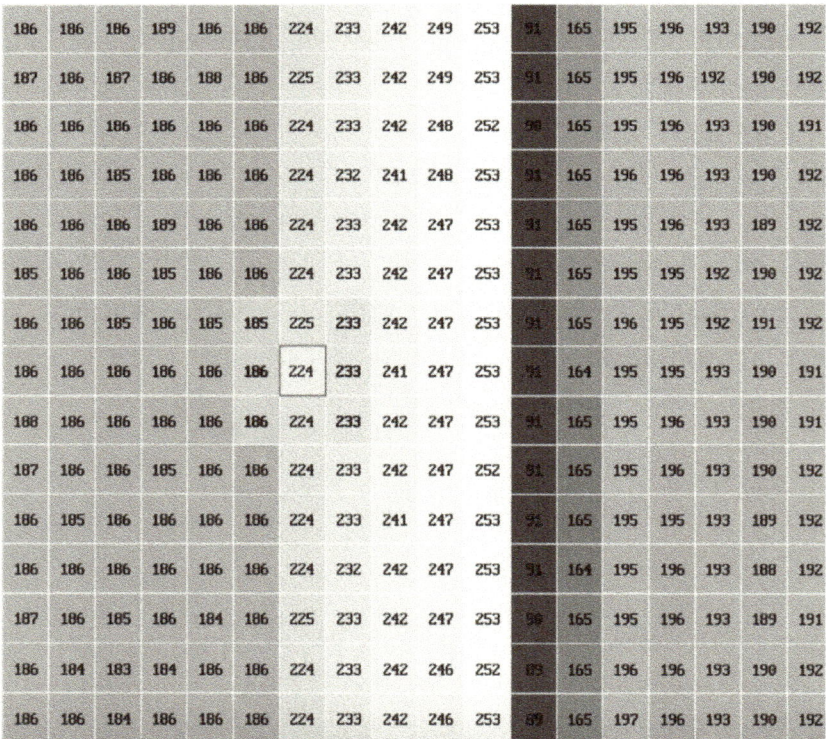

186	186	186	189	186	186	224	233	242	249	253	91	165	195	196	193	190	192
187	186	187	186	188	186	225	233	242	249	253	91	165	195	196	192	190	192
186	186	186	186	186	186	224	233	242	248	252	90	165	195	196	193	190	191
186	186	185	186	186	186	224	232	241	248	253	91	165	196	196	193	190	192
186	186	186	189	186	186	224	233	242	247	253	91	165	195	196	193	189	192
185	186	186	185	186	186	224	233	242	247	253	91	165	195	195	192	190	192
186	186	185	186	185	185	225	233	242	247	253	91	165	196	195	192	191	192
186	186	186	186	186	186	224	233	241	247	253	91	164	195	195	193	190	191
188	186	186	186	186	186	224	233	242	247	253	91	165	195	196	193	190	191
187	186	186	185	186	186	224	233	242	247	252	81	165	195	196	193	190	192
186	185	186	186	186	186	224	233	241	247	253	91	165	195	195	193	189	192
186	186	186	186	186	186	224	232	242	247	253	91	164	195	196	193	188	192
187	186	185	186	184	186	225	233	242	247	253	90	165	195	196	193	189	191
186	184	183	184	186	186	224	233	242	246	252	83	165	196	196	193	190	192
186	186	184	186	186	186	224	233	242	246	253	89	165	197	196	193	190	192

Figure 8.10: Pillar's edge

Source: Author

-1	0	1
-1	0	1
-1	0	1

Figure 8.11: Pillar's edge

Source: Author

This operation of applying kernel to a patch of pixel is called a convolution.

185 ×-1	225 ×0	233 ×1
186 ×-1	224 ×0	233 ×1
186 ×-1	224 ×0	233 ×1

Figure 8.12: convolution

Source: Author

These edge-enhancing kernels are called Prewitt operators. Sharpening and blurring an image both have their own kernels. These types of kernels begin to characterize simple shapes. For example, the bridge of the nose on a face tends to be brighter than the side of the nose. Eyes, on the other hand, look more distinctive, with dark circles surrounded by lighter patterns.

Although each kernel is a weak face detector by itself, they can be quite accurate when they're combined. This was the basis of an early algorithm called Viola-Jones Face Detection.[14] Today, a convolutional neural network replaces it. Rather than one-dimensional input, we pass a neuron 2D pixel data, and the input weight corresponds to the kernel value, in which the kernel learned its own.

For example, when an image is passed through the banks of neurons to process data, the first few convolutions may do a simple task like finding an edge. The next layer of convolution might find shapes like the eyes and mouth. A layer beyond that might be responsible for finding corners. After many convolutions, the AI can detect a face.

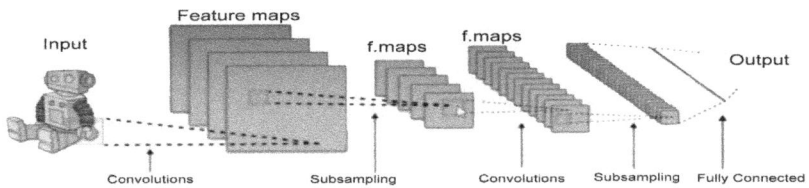

Figure 8.13: Computer Vision

Source: Author

Natural Language Processing

At the beginning of this chapter, we talked about how Alan Turing proposed the Turing Test for machine intelligence. In order to pass the Turning test, a computer must be able to impersonate a human so well that a human judge cannot distinguish it from another human. The skillful use of language is a major part that makes us human. And for this reason, the desire for computers to understand and speak our language has been around since they were first conceived. The technique is another important field of AI, which is called Natural Language Processing. (NLP).

NLP mainly explores two big ideas: natural language understanding and natural language generation.[15]

Natural language understanding refers to AI that filters your spam mail or instructs your self-driving car on how to get to a friend's house. Natural language generation, on the other hand, is AI that performs translation of a document, summarizes text or chats with you. The key to both natural language understanding and natural language generation is to understand the meaning of a word, which can be tricky sometimes because words have different meanings in different contexts. Say, for example, I say, "Meet me at the bank." It can have two meanings. First, the bank can mean the actual bank to withdraw money. Second, it could mean the riverbank.

Chatbots is a computer program that simulates human conversation through voice commands, text chats or both.[16] In the 90s, when I was a kid, my grandfather showed me a computer program called ALICE, the first well-known chatbot that people could interact with online.[17] We were fascinated by how we could actually talk to a computer! Fast forward to today, and chatbots have become part of our daily life. The modern chatbots approach is based on machine learning, where gigabytes of real human-human chats are used to train chatbots. Today, chatbots are used in customer service, where there are heaps of conversations to learn from.

In 2017, Facebook began an experiment on two chatbots negotiating with each other. The robots, nicknamed Bob and Alice, were originally communicating in English. As the conversation goes on, they swapped to what initially appeared to be gibberish.[18] Eventually, the Facebook researchers that control the AI realised that Bob and Alice had, in fact, developed their very own, seemingly more efficient language. In the end, Facebook engineers had to shut down the AI system after the chatbots began talking to each other in a new language that humans couldn't understand.

It is not only just Facebook. In 2016, Google Translate used a neural network to translate between popular languages and between language pairs for which it has not been specifically trained. It was revealed that Google Translate used its own sort of "intermediary" language to translate between a pair of languages to which it hadn't been exposed. To do that, Google Translate seemed to have developed its own artificial language.[19]

Swarm Intelligence

Ants don't have traffic jams. Even hundreds of thousands of ants walk along narrow lanes, in and out of their nest, wouldn't bump into each other. Ants, birds, fish all exhibit some knowledge of swarm intelligence and perfectly synchronize their movements.

Long ago, scientists already realized that complex patterns could arise from simple individual behaviors in a flock of swallows or an ant colony. The science of swarm behavior inspired scientists to build robots of the future – one that could exhibit swarm intelligence from building construction to search-and-rescue missions. What makes swarm intelligence so perfect for robots is that not one robot in a swarm does anything too complicated. The best thing about swarm intelligence is that you don't need to tell each robot exactly what to do. Instead, you give the robots a bunch of the same basic rules, and because of how those rules play out in large numbers, the group will self-organize and figure out how to do the complicated task you want them to do.

In 2014, Harvard researchers had developed a simple, low-cost robot called Kilobots that used vibration motors for locomotion and infrared light for communication. The team mass-produced 1024 robot swarms for testing collective behaviors.[20] They designed an algorithm that allowed robots to robustly form the desired shape without human intervention. The researchers never told the robots where to go; they just gave them a set of basic rules to follow, like measuring how far from neighbor robots and finding an edge and moving along it; the effect is similar to a flock of birds wheeling across the sky.

The Harvard researchers also took inspiration from termites to make robots able to build. Stigmergy is the strategy that learns from termites, which indirectly communicates with each other to achieve a common goal.[21] For humans, when we construct, we need blueprints and checklists, which involves a lot of direct communication. Termites build by paying attention to clues left behind by fellow termites in their environment. So, each termite is doing its own thing, but they indirectly coordinate with one another as well. This technique from termites can train robots to learn how to build. During construction, each robot will place its block based on how the previous block was put down and orientated.

In the future, it is entirely possible to use swarm robots to assist in construction in dangerous places not possible for humans to build (e.g., the lunar base or Mars).

Apart from that, scientists discovered that traffic could flow more smoothly if we apply swarm intelligence in the autonomous vehicles of the future. If you drive to work, you will know what I mean by that. We are pretty bad at communicating well on the road. Swarm intelligence can be used to train cars to stay at a distance from one another.

Can AI Take over the Future?

In January 2015, Stephen Hawking, Elon Musk, and dozens of artificial intelligence experts signed an open letter on artificial intelligence, calling for research on the societal impacts of AI. The letter affirmed that AI could deliver incalculable benefit to society. It has the potential to eradicate disease and poverty, but at the same time, it could potentially end the human race if deployed incautiously.

In general, there are two types of artificial intelligence: narrow AI and general AI.

The AI we discussed in this chapter are examples of narrow AI. Although deep learning in narrow AI is very powerful, in reality, the narrow AI we have today can only do one thing at a time. AlphaGo might have defeated the world's chess champion, but it is only good at playing chess; it is terrible at playing music. Siri is good at NLP but probably not good at computer vision. These AI are called weak AI or narrow AI. They are only good at one specific task. It is this type of AI that is displacing people from jobs today. It is doing specific jobs blue collars and white collars are paid to do.

Artificial General Intelligence (AGI), on the other hand, is different from narrow AI. General AI can reason, solve puzzles, make judgments, plan, learn, conceive new ideas and communicate. It is this type of AI that the experts in the field worry about. Prof Hawking says the primitive forms of artificial intelligence developed so far have already proved very useful, but he fears the consequences of creating something that can match or surpass humans. In reality, we cannot quite know what will happen if a machine exceeds our own intelligence and coexists with us. We are not sure if this will ever happen.

The history of AI has undergone boom and bust cycles. During the cycle, people overhype and made wild claims about AI. Eventually, when disappointment inevitably follows, it caused the public to withdraw funding into the AI industry, which leads to AI winter - a period of reduced funding and interest in artificial intelligence research.[22] Right now, we are in AI spring. It is entirely possible we could be heading into another AI winter. But, this time, we are entering the AI twilight zone between narrow and general AI. Experts in the industry believe we are at a turning point in AI history. Four surveys with 995 AI experts suggest the emergence of AGI or the singularity by the year 2060.[23]

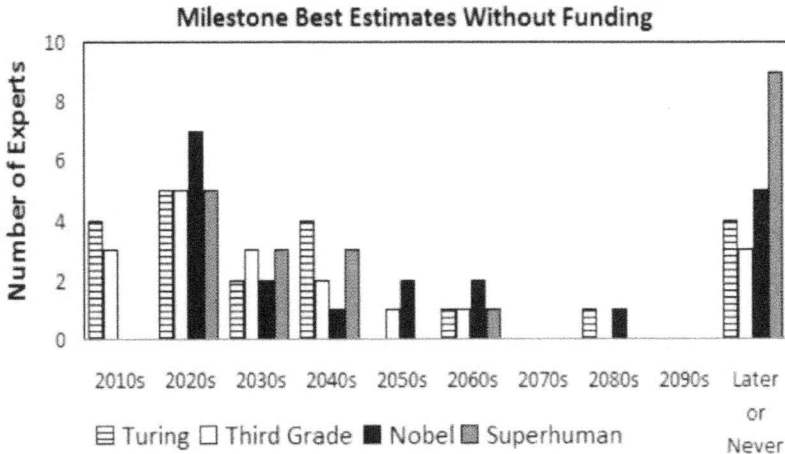

Figure 8.14: Milestone Best estimates Without funding

Source: Survey distributed to attendees of the Artificial General Intelligence 2009 (AGI-09) conference

[Note: Milestone best estimate guesses without massive additional funding. Estimates are for when AI would achieve four milestones: the Turing Test (horizontal lines), third grade (white), Nobel-quality work (black), and superhuman capability (grey).]

***

AI is a constellation of technologies. Today, the potential of AI has moved away from mere automation of simple tasks. It transforms or redefines the relationship between people and machines. It is an extension of our capability, allowing us to make better, faster and precise decisions.

Even so, the future of AI promises a new era of both promises and dilemma. After all, how could humans stay in control of a complex intelligence system? This poses a serious question about artificial intelligence: will it have the same advantage over us one day?

When that day comes, this is when some expert called singularity – a point in time where human beings are no longer the most intelligent beings on Earth.

Chapter 9
Doctor in A Cell

*F*antastic Voyage is a 1966 American science fiction film. In the film, a brilliant scientist, Jan Benes, had developed a way to shrink humans and objects for a brief period of time. Benes, who worked for communist Russia, was transported by the CIA to America but was attacked. To save the scientist, who had developed a blood clot in his brain, a team of Americans in a nuclear submarine were shrunk and injected into Bene's body. They had a finite period of time to remove the blood clot and escape before the miniaturization wore off.

While *Fantastic Voyage* may just be a movie, in the future, researchers will be able to create a programmable biological computer that may navigate within the human body to diagnose disease by combining computer science and molecular biology. This is what Professor Ehud Shapiro coined it: *Doctor in a cell.*[1]

An Eternal Battle

Humans have a deep history of viral infection. Cholera, the Black Death, smallpox and influenza are some of the most brutal killers in human history. In the early 20th century, something unexpected happened that revolutionized the field of medicine today.

On 28th September 1928, Alexander Fleming, a Scottish researcher, began to sort through Petri dishes containing colonies of bacteria. He noticed that something unusual had happened in one of the dishes. On one dish, he found mold growth, but in the area around the mold growth, there were no bacteria. There was when Fleming discovered penicillin – an accidental discovery that changed the world of medicine.[2]

We called this antibiotic.

Prior to the beginning of the 20th century, infectious diseases caused high mortality worldwide. The average life expectancy at birth was only 47 years old. The discovery of antibiotics is a singularity in the field of medicine. It revolutionized the treatment of infectious disease worldwide.

However, our victory against bacterial infection seemed to be short-lived. A growing threat of antibiotic resistance – the ability of bacteria to resist the effect of antibiotic is threatening global health and food security.[3]

The threat of antibiotic resistance of bacteria is real. It can affect anyone, any age, and country. A growing number of infections, such as tuberculosis, is becoming harder to treat as the antibiotic used against them is becoming less effective.

We are fighting a losing battle against it…

Fortunately, a futuristic way to treat disease is on the horizon, which will revolutionize how disease will be treated. Before we go into that, let's dive inside the microscopic world at how we get sick in the first place?

The Biological Computer

The human body is made up of 37.2 trillions of cell. These cells work in harmony to carry out their basic function to keep us alive.

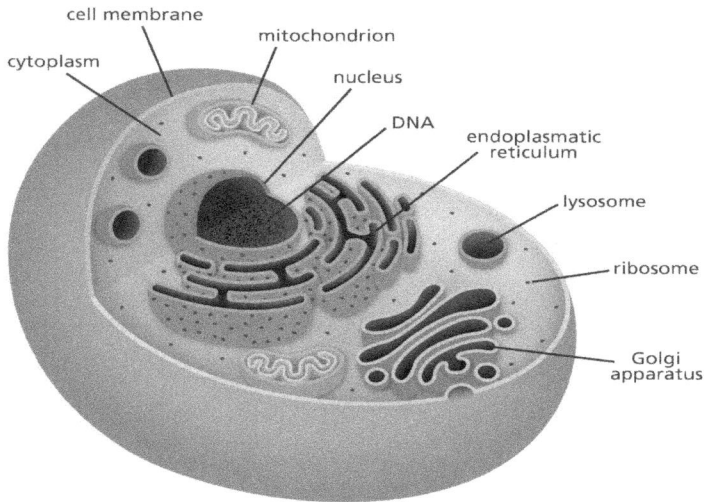

Figure 9.1:Anatomy of a cell

Source: yourgenome.org

In the nucleus of each cell are 23 pairs of chromosomes, a total of 46 chromosomes. Inside the chromosomes are our genes, which are long, double helical molecules composed of bases. We call them deoxyribonucleic acid (DNA). DNA is the blueprint of life. The order of these bases determines the genetic blueprint. It stores genetic information like our height, eye color, blood type and color etc. These traits are passed on from parents to children through genes.

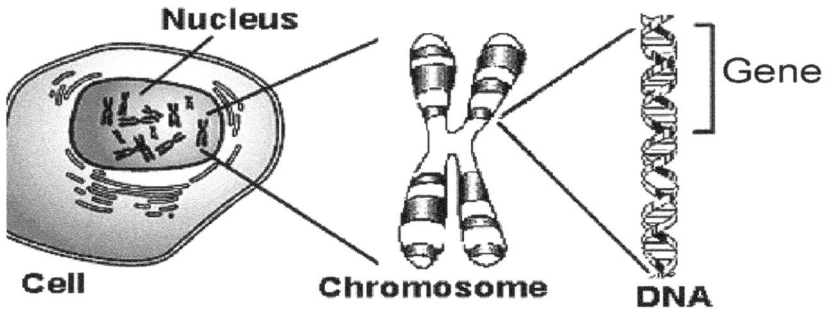

Figure 9.2: Cell ,chromosome and DNA

Source: Author

[Note: A nucleotide is the basic structure unit and building block of DNA. These building blocks hook up together to form a chain of DNA. It composes a five-side sugar, phosphate group and nitrogenous base. The sugar and phosphate group make up the backbone of the DNA helix structure, with the bases located in the middle. Bases are the part of a DNA that holds information. There are four bases in DNA called Adenine (A), Cytosine (C), Guanine (G) and Thymine (T).]

To read this DNA blueprint, the DNA is unzipped to expose individual strands, and an enzyme translates them into intermediate message Ribonucleic acid (mRNA). This mRNA will then carry the instruction outside the nucleus to a molecular machine called ribosome to manufacture proteins.

If you take a step back, you will realize that every cell in our body is like a tiny factory. Our DNA directs every activity to constantly producing proteins. The whole process of protein manufacturing and DNA replication mechanism is like the function of molecular machines.

[Note: Protein, in the context of biochemistry, is a lot more than what is for dinner. Enzyme, antibodies, and even our muscles are all proteins.]

How Do We Get Sick?

Besides bacterial infection, another most common infectious disease that makes us sick is viral infection.

Viruses enter our body through the air we breathe and the food we eat. When a person coughs in his hand and touches a doorknob, he places the virus on the doorknob. The next person in contact with the doorknob will pick up the virus. If that person then touches his face or nose with the unwashed hand, the virus will be inside the body.

Unlike bacteria cells, viruses are something in between living thing and non-living thing. They need a cell to reproduce themselves. When a virus makes contact with a cell's membrane, it inserts its genetic coding into the cell. The cell absorbs the viral genome into its own genetic materials, and the virus takes control of the cell functions. The cell begins to produce offspring of the original virus. The new viruses are then released from the cell and infect neighboring cells. This is how we get sick.

Coronavirus (COVID-19)

During December 2019, a novel coronavirus virus out broke in Wuhan, China. The virus was rumored to start in the Huanan Seafood wholesale market in Wuhan. Soon, it spread to the rest of the world.[4] Those who are infected may develop mild to severe symptoms. Some asymptomatic persons might have no symptoms at all.

As I am writing right now, the total cumulative confirmed cases is 32,807,114 and 994,838 deaths from the virus. The incubation period can be up to 14 days or more. It is a global epidemic, and worst of all, there is no cure for it.

Medicine of the 21st Century

While lethal biological viruses pose a risk to the human race, researchers learned how viruses deliver their genetic material into human cells. And by simply switching the material a virus uploads into cells by removing its genes and inserting therapeutic ones, researchers use viral vector therapy that engineered virus to deliver genes into cells. Viral vectors like adeno- associated virus (AAV) are especially good at treating retinal diseases. This works by eliciting a weak immune response.[5]

Nanorobotics is an emerging technology in the 21st century, creating machines close to the scale of nanometer. The first useful application of nanorobotics is nanomedicine. To give us an idea of how small nanoscale is, nanoparticles are somewhere between atoms and molecules. A nanoparticle is about 5nm in length, and a virus is about 50nm. We simply cannot see nanoscale with our naked eye. Nanotechnology is the science, engineering and technology about 1 to 100 nanometers.

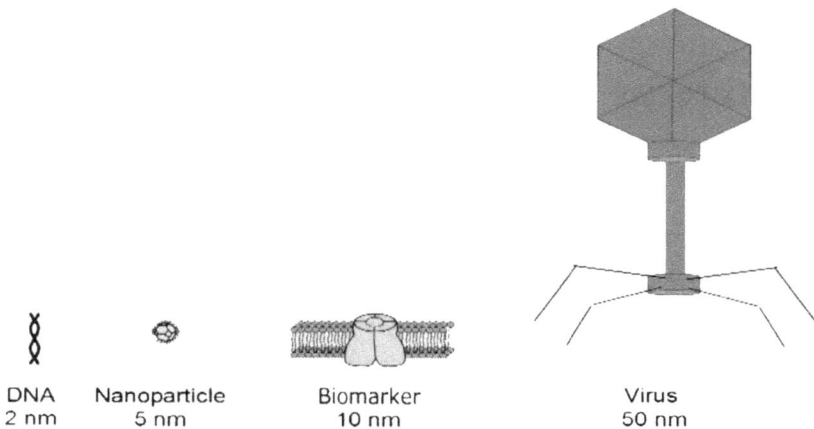

| DNA | Nanoparticle | Biomarker | Virus |
| 2 nm | 5 nm | 10 nm | 50 nm |

Figure 9.3: size of DNA, nanoparticles, biomarker and virus

Source: Author

In 1959, physicist Richard Feynman said in his speech, *"There's Plenty of Room at the Bottom*," that one day, machines would be miniaturized, and huge amounts of information could be encoded in minuscule spaces.[6] In 1986, K. Eric Drexler's book, *Engines of Creation: The coming era of nanotechnology*, reinforced this idea. In his book, Drexler posited the idea of programmable self-replicating nanomachines.[7] Because these nanomachines are programmable, you can simply control them and build things at an atomic level, just like how kids play with LEGO bricks. While this idea might sound like science fiction, in reality, we cannot discount the idea that we are alive today because of countless nanobots operating within each of our trillion cells. Like ribosomes (25nm), some of these nanobots are programmed with functions to take information encoded in mRNA and translate into a specific sequence of amino acids.

But how are nanorobots created?

Obviously, we cannot build machines at nanoscale using conventional engineering tools. Things in nanoscale follow a more complex law of quantum mechanics. We need a different set of tools to manipulate atoms and molecules in a way to use them as building blocks for nanomachines.

In 2016, a group of German physicists built the smallest engine ever created from just a single electrically charged calcium atom. Like all other heat exchange engines, this machine converts thermal energy into mechanical movements. Its atom is heated by electrical noise and cooled by shooting a 397nm laser beam. By switching between cooling and heating, the atom is moved back and forth, following the heating and cooling cycles like a heat exchange engine.[8]

Mechanical engineers at Ohio State University have designed and constructed complex nanoscale mechanical parts using a method called

DNA origami – a technique that uses the nanoscale folding of a long DNA strand to create any non-arbitrary two or three-dimensional shape at the nanoscale[9],and then securing certain parts together with staples made from shorter DNA strands. You can think of it as DNA Lego, connecting together at different angles.

Jørgen Kjems, Kurt Gothelf and colleagues from Aarhus University, Demark, used DNA origami to make a box shape.[10] To do this, the team took a long, circular single-strand DNA from a virus that infects bacteria called bacteriophage M13. The team used a computer to work out the exact combination of short strands of complementary DNA that could staple the appropriate area of the ring to get the desired box shape. By adding extra sections of DNA to the right staple strand, they formed the 'locks' on the rim of the box. The team attached fluorescent molecules to the two parts of the lid. So, when the box was closed, they got a red fluorescent signal. When it was opened, the signal turned green. This is a milestone as the box can have different locks and open or close in response to different things. Kjems revealed that the box could be quite big – 30nm.[11] His team had success in putting cargoes, such as virus, enzyme or macromolecules inside the box. Kjems can foresee two of the applications are the control release of drugs and sensors. DNA origami is going to be a widely used method for making nano-structure, and a standard procedure for many labs developing future drug delivery systems and electronics.

The most exciting part of using nanobots for drug delivery is that it is possible to get drugs to its destination faster and more accurately. It does not have to rely on the body's natural circulatory system to get medicines where it needs to go.[12] This can reduce the side effect of powerful drugs. In some particularly impressive studies, researchers attach drugs to bacteria and use magnetic fields to guide them to tumors.

Putting a synthetic nanomotor with the correct fuel, a nanomachine will move continuously in our body noninvasively.[13] By attaching a bio receptor to a constantly moving nanomotor, it will collide with target molecules faster than simply floating randomly.

CRISPR-Cas9

Bacteria and viruses have been fighting since the dawn of life. Certain viruses need to infect bacteria to reproduce. On the other hand, bacteria do not want to be infected.

Viruses infect bacteria by inserting their own genetic material into the bacteria. Most bacteria that get infected by viruses will die. Every so often, a lucky bacterium survives viral infection because of a mutation in that bacterium's DNA. Mutations are like little mistakes. They are changes to the DNA sequence of a gene. While some of these mistakes kill the bacterium, others actually help it to fight off the virus. If they do so through helpful mutation, they activate their most effective antivirus system. They save part of the virus DNA in their own genetic code in a DNA archive called Clustered Regularly InterSpaced Palindromic Repeats (CRISPR). When the virus attacks again, the bacterium quickly makes an RNA copy from the DNA archive and, therefore, arms a secret weapon – a protein called CAS9. The protein scans the bacterium's inside for signs of the virus invader by comparing every bit of DNA to the sample from the DNA archive.[14] When it finds a 100% match, it is activated and cuts out the virus DNA, making it useless, thus, protecting the bacterium against the attack. CAS9 is very precise, almost like a DNA surgeon.

CRISPR is a description of some special region in a bacterium DNA.

At this sequence, there are two types of DNA sequences that alternate: repeats and spacers.[15] Repeats (black diamonds) are the letters that repeat over and over. Spacers (colored squares) are in-between repeats and are all different. When scientists first found these spacers, they were not sure what they were for. But then they discovered that these spacers are very similar to virus DNA.

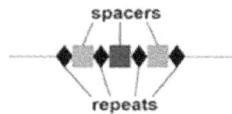

Figure 9.4 : Spacers and repeats in CRISPR regions

Source: Author

While humans are not bacteria, scientists are learning the same CRISPR technique to fight virus. The revolution began when scientists figured out that the CRISPR system is programmable.

In 2007, Rodolphe Barrangou explored the idea of whether CRISPR could possibly help bacteria to recognize and fight viruses. He and his team compared the CRISPR region of non-virus-resistant type of bacteria to the virus resistant type of the same species. What they found out is that the virus-resistant version had some extra spacers. These spacers were very similar to the virus that they were exposed to. Barrangou and his lab deleted and inserted several spacers that matched the same virus. They found out that when they deleted a spacer from a virus-resistance bacterium, the bacterium lost its resistance to the matching virus. When they added back the spacers, the bacterium regained its resistance again. So, Barrangou concluded that these spacers in the CRISPR regions provide resistance to viruses by saving some of the virus' DNA, thus, allowing the bacterium to remember them.[16]

By using this knowledge, scientists figured out how CRISPR works to prevent viruses from destroying the cells. The bacterium clips a bit of the viral DNA and adds it into a CRISPR region of its own DNA. So, if the virus ever comes back, the bacterium makes RNA from the region of CRISPR specific for that virus. These RNA copies pair up with some CRISPR-associated proteins (cas9). The RNA guides the cas9 protein to the invading viral DNA, so the protein can destroy it. No more viral DNA, no new viruses.

Cancer Could Be a Disease of the Past

Cancer remains a leading cause of death worldwide. Unlike a virus, one of the main reasons the human body cannot fight cancer is because it cannot recognize it as a foreign agent. A cancer cell has the patient's DNA. As the body might not produce enough T cells to fight cancers, the immune system fails. In some cases, some cancers can weaken the immune system by spreading into the bone marrow that makes white blood cells to fight the infection.

Traditional chemotherapy uses drugs to destroy cancer cells. It works to keep cancer cells from growing, diving and making more cells. But chemotherapy treatment cannot differentiate between normal cells and cancer cells. It kills the healthy cells in the process.[17]

CAR T-cell (Chimeric Antigen Receptors T-cells) therapy is very different from traditional treatment. This immunotherapy uses the body's own immune system or components of the immune system to fight cancer. In the laboratory, a modified virus containing a CAR sequence is used as a viral vector, which can be Lentivirus or retrovirus, transferring targeted genetic information into normal T-cells incorporated into the genome. The final result is a T-cell, expressing the desired CAR receptor.

These CAR T-cells will be expanded, grow in large numbers and infused back to the patient. The CAR T-cells receptors will bind specific antigens expressed by the target cancer cells, killing them.[18]

In 2010, at the Children's Hospital of Pennsylvania, a 5-year-old girl was diagnosed with acute lymphoblastic leukaemia (ALL), and despite the many cycles of chemotherapies, she experienced several relapses. Out of options and a fatal prognosis, one oncologist made the parents aware of a new experimental therapy. The parents opted to enroll her in this new study, and she became the first child to receive chimeric antigen receptor T-cells (CAR-T) cells infusion. Now, she is only 12-year-old, but this success helped reverse that line of research that was close to failing[19], and, in 2017, the FDA approved the first CAR-T cell technology.[20]

Although CAR-T cell is one of the most promising ways to counter cancer, it has its own limitations. In some patients, CAR-T cells fail because the tumor cells multiply so quickly that it sheds that target CD-19 molecules on the surface of the cancer cells. Tumors are able to bypass the immune system and spread from the original growth site. Tumor cells are able to highly express the programmed death-ligand 1 (PD-L1), which binds to programmed death-1 (PD-1) receptors on T-cells and inhibits their effectors and function.[21] Also, tumor cells can secrete cytokines that inhibit the proliferation of cytotoxic T cells. So, to be effective, it is indeed necessary to overcome tumor-induced immune suppression and suppress the tumor microenvironment (TME).[22] The CAR T-cell will need to recognize two or more different molecules on the cancer cell, even if the target one disappeared.

[Note: CD19 is B-lymphocyte antigen, used to diagnose cancers arising from B-lymphocyte.[23]]

Apart from CAR T-cell therapy, target drug delivery is another possible alternative for cancer treatment. A liposome is a tiny bubble (vesicle), made out of the same material as a cell membrane. Liposomes can be filled with drugs and used to deliver drugs for cancer and other diseases. A fluorescent semiconductor, known as "quantum dots," is small particle that glows in the dark by coating it with nano gold particles.[24] They are useful because when they are attached to a cell of a molecule, they behave like a beacon, so we can see them. Magnetic metal oxides are particles that respond to magnets. They can draw fluid with it to the magnet. By using a magnet as a guide, magnetic metal oxide can be controlled at a nanoscale, non-invasively, using a magnetic field.[25]

Another technique of eliminating cancer cells is by using CRISPR. CRISPR gives us the means to edit the immune cells and make them better CRISPR hunters. So, in the not too distant future, you can get a couple of injections of a few thousand of your own immune cells. The first clinical trial for a CRISPR cancer treatment on human patients was approved in early 2016 in the U.S.[26] In less than one month, Chinese scientists announced that they could treat lung cancer patients using immune cells modified with CRISPR in August 2016.[27]

Personalized Medicine

Drug discovery takes a long time. It takes 3-6 years for drug discovery to preclinical, 5 years for clinical trials, and another 2 years for review. By the time a drug gets FDA approval, it will be very expensive. According to a recent study by Tufts Center for the Study of Drug Development and published in the Journal of Health Economics, developing a new prescription medicine that gains marketing approval is estimated to cost drug makers $2.6 billion.[28]

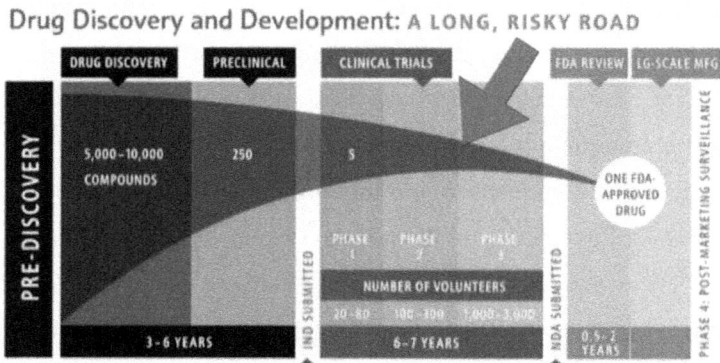

Figure 9.5: Pharmaceutical research and manufacturers of America

Source: pharmaceutical research and manufacturers of America

Imagine if there is a way to bypass years of expensive initial research, lab testing, human testing and multiple phases of patient testing? Imagine if the treatment is designed to target your individual genome from lab to clinical trial at a lower cost?

A new approach in medicine is needed.

This is where personalized medicine comes into play.

Since the first human genome was sequenced in 2003 [29], the cost of genome sequencing drops exponentially. From sequencing the genomes of newborn babies to conducting national population genomics programs, the field is gaining momentum and getting more personal by the day. New sequencing systems, like the DNBSEQ-T7 from the BGI Group, the world's largest genomics research group, are pushing the technology into broad use.[30] The system generates 60 genomes per day, equaling 6 terabytes of data. NVDIA Tensor core GPU accelerates genomics sequencing through AI deep learning.[31]

Wyss Institute researchers have adapted a computer microchip manufacturing method to engineer a microfluidic culture device that recapitulates the microarchitecture and functions of living organs like our lungs, liver, etc.[32] These microdevices called Organ-On-chips offer a potential alternative to traditional animal testing. Each Organ Chip is composed of a clear, flexible polymer, about the size of a computer memory stick. It contains hollow microfluidic channels lined by living human organ-specific cells, interfaced with a human endothelial cell-lined artificial vasculature. Mechanical forces can be applied to mimic the physical microenvironment of living organs, including breathing motions in lung and peristalsis-like deformations in the intestine. They are essentially living, three-dimensional cross-sections of major functional units of whole living organs.

In the future, Organ-On-chips could be tailored to your genetic makeup. These chips will be used to test treatment until one fits you, thus bypassing the expensive, lengthy FDA approval process.[33] The Biotech company, Emulate, has raised millions to use organ-on-chips to make personalized medicine a reality.[34]

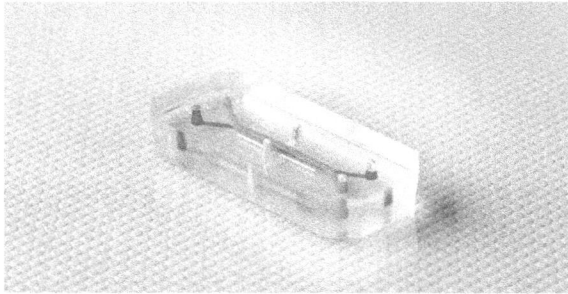

Figure 9.6: Organ on a chip

Source: pharmaceutical-technology.com

Bowhead Health, a health data company, uses an alternative approach in personalized medicine. Bowhead's biometric home device kit reads biometric data, like saliva or a blood prick test, in real time and transmits the reading to your doctors. As soon as the analysis is completed, and your deficiencies are identified, your home device kit will recommend a customized vitamin-based pill for you.[35]

The AI Doctor is ready to see you

When MIT Professor Regina Barzilay received her breast cancer diagnosis, she turned her own case into a science project. Knowing the disease could have been detected earlier if the doctors were meticulous enough to recognize the signs beforehand, Barzilay used a collection of 90,000 breast cancer x-rays to develop a program that could predict a patient's cancer risk.[36]

With deep learning and machine learning, the AI doctor evaluates huge amounts of data from databases to make accurate predictions as well as recommend interventions. AI is able to pick up small details, which human eyes cannot pick up. It will be better than human doctors at diagnosing diseases like cancer. As machine learning becomes better with more data, AI can revolutionize the future of medicine. It might even allow us to see the future of our bodies. But, as good as this might sound, the quality of the databases must be prioritized. If bad data is implemented, AI will make the wrong recommendation and put the patient at risk.

CheXNet, a 121-layer convolutional neural network, inputs a chest X-ray image and outputs the probability of pneumonia along with a heatmap localizing the area of the image most indicative of pneumonia. When scored against other radiologists' performance on the pneumonia detection task, it was found that CheXNet achieves an F1 score of 0.435 (95% CI 0.387, 0.481) than radiologists 0.387 (95% CI 0.330, 0.442).[37] AI algorithm is getting really good at using deep learning to diagnose diseases in X-rays.

Merging Your Brain with Machines

When you hear the word cyborg, you would probably remember the famous movie Robocop in 1987. While the heroic characters in the film seem futuristic today, it might no longer be mere science fiction in the future.

We are rapidly moving towards an era where artificial intelligence, digital technology and human biology are converging. Humans are already able to control robotic arms with their minds.

Today, we have prosthetic limbs tied to a person's nervous system. Tomorrow, we will see computers wired to our brains to enhance our ability.

Neuralink Corporation is an American neurotechnology company founded by Elon Musk in 2016. *Neuralink*'s goal is to develop implantable brain-machine interfaces.[38] In July 2019, Elon Musk made a presentation detailing the *Neuralink* project – a venture that merged human with AI.[39] Human cognition has two major systems – the limbic system and the cortex system. The limbic system is involved in motivation, emotion, learning, and memory. The cortex system, on the other hand, plays a key role in attention, perception, awareness, thought, memory, language, and consciousness. *Neuralink*'s mission is to implant a third system that augments humanity with computers and eventually, artificial intelligence. Depending on how you look at it, we already have this layer in the form of our phones and laptop. We can interface with information at our fingertips and speech. But these methods of communication are too slow and cost too much bandwidth. A much faster approach is needed so that we can communicate with our device at a much faster rate. This is where the brain-machine interface (BMI) becomes the focus of *Neuralink*.[40]

Our brain consists of 86 billion neurons, firing electrical signals all the time whenever we talk or think. Neurons are essentially electrical devices. At the junction between neurons is the synapse. Neurons communicate with the synapse, using chemical signals called chemical neurotransmitters. The process by which neurons send signals is called action potential, which is a spike or impulse. Action potential produces an electrical field spread from neurons, which can be detected by placing an electrode. The goal of Neuralink is to record and selectively stimulate as many neurons as possible.[41]

I know we all have concerns about placing an electrode inside the brain.

The electrode designed by Neuralink is non-invasive. They are tiny threads about 1/10 the cross-sectional area of human hair. These threads, measured to be about 24 microns, would be inserted individually into the brain with micron precision by neurosurgical robots. This would be able to target specific parts of the brain without puncturing any blood vessels. In the future, Neuralink hope to use laser beams to pierce through tiny holes in the skull that wouldn't be felt by patients. [42] It would be as safe and painless as laser eye surgery - LASIK.

The future of Neuralink will be in three stages. Stage one will be to understand and treat brain disorders, starting with people with serious medical needs. It allows disabled people to control electronic devices or move robotic arms using only the brain as the input signal. Stage two will be to preserve and enhance one's brain. In stage three, we will have a full brain-machine interface, which can be a kind of app store or program that you can download and control your brain. Other possibilities include a new kind of communication, like telegraphy or downloading memory of someone familiar with a city so that you will be familiar with it when you go to a city. The possibilities are endless.[43]

Road to Longevity

We want to fix the things we don't like about the changes that happen between the age of 30 and the age of 70.

- Aubrey De Grey

Longevity has been a dream by kings. Long living is a good wish or blessing in Asian culture. Scientists have been constantly extending the human lifespan. From 1770 until 2019, our life expectancy increased from 29 to 73. Life expectancy increased in all countries around the world. Global record life expectancy from 1840 to today increased by one year every four years.[44]

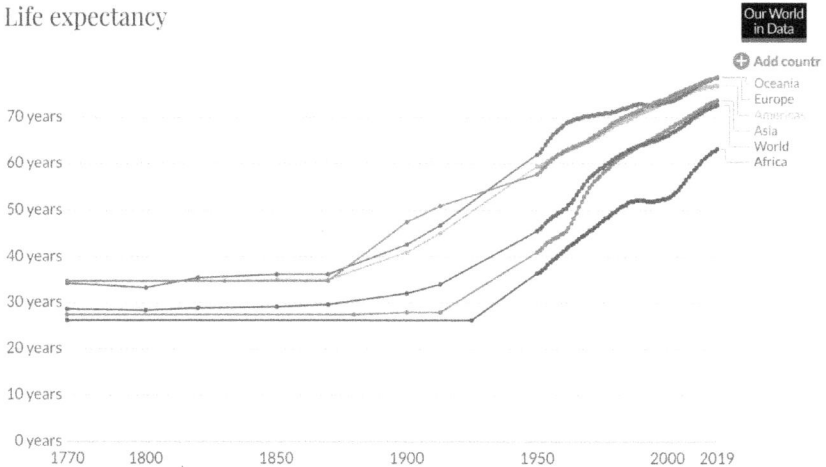

Figure 9.7: Life expectancy

Source: Railey(2005) Clio Infra (2015), and UN Population Division (2019)

But if you think about it, what we are experiencing today as natural actually wasn't natural at all to our ancestor. The medical evolution in the past 250 years has increased our life expectancy by 2.5 times. Even so, today, disease and biological aging remains our roadblock to immortality.

While technological, immunological and nanotechnology might be a possible solution to severe disease we see today, how about aging?

For human and animals, aging is a process that happens over the course of our lives. The process of aging has a name called *Senescence*.[45] At an organism level, our Senescence takes place after we have reached sexual maturity. Afterwards, we begin to lose our ability to combat stress, maintain homeostasis, the internal balance of all our organs and our ability to combat disease.

Strangely enough, not all animals age in the same way. Some of the cold-blooded animals exhibit negligible *Senescence*. They don't lose their ability to reproduce over time, and their death rate does not increase with age either.[46] They died because of disease or something terrible happened to them.

The Galapagos tortoise has been known to live up to the age of 170.[47] Lobsters, on the other hand, can live up to 140 years old.[48] The oldest quahog clam lived up to a record age of 405 years old![49]

So why do some animals start to die sooner, but some can live for up to four centuries?

Scientists try to figure this out by analyzing senescence at a cellular level.

It turns out that our cells have death programmed in them.

Our somatic cells constantly divide and make copies of themselves.

In 1960, the young scientist, Leonard Hayflick, was studying human fetal cells. He noticed that human cells just stop dividing when they died. He observed that cells quit dividing after about 50 divisions, which took about 9 months. If you put your healthy dividing cells into a freezer, the cell division would slow down or even stop. But when you warm them up again, they will pick up where they left off and start dividing again until they reached 50 divisions. Hayflick realized that there is *indeed death programmed in our cells*. The number of times a cell can divide is called the *Hayflick limit*.[50]

Although human fetal cells have about 50 divisions before they die [51], not all animals are the same. Mice can live for 2 to 3 years with 13 to 28 divisions. The Galapagos tortoise, on the other hand, has a Hayflick limit of about 125. Therefore, there must be some correlation between an animal's Hayflick limit and its life span.

In reality, it is not that simple. As we get old, our Hayflick limit changes. Studies of people in their 80s or 90s found that the cells of people in this age group only show about 20 cell divisions.[52] But, what causes that limit to change as we get older?

The answer might be found in chromosomes.

A human cell stops dividing after it stops being able to completely replicate its telomeres – the little cap of non-coding DNA that protects genes from error of copying. Telomeres are originally made with the help of enzyme Telomerase.[53] But, after every cell division, the telomeres of chromosome of new cells are a tiny bit shorter. A Hayflick limit is reached when the telomere is so short that they can no longer protect the chromosome, and the cell becomes unviable.

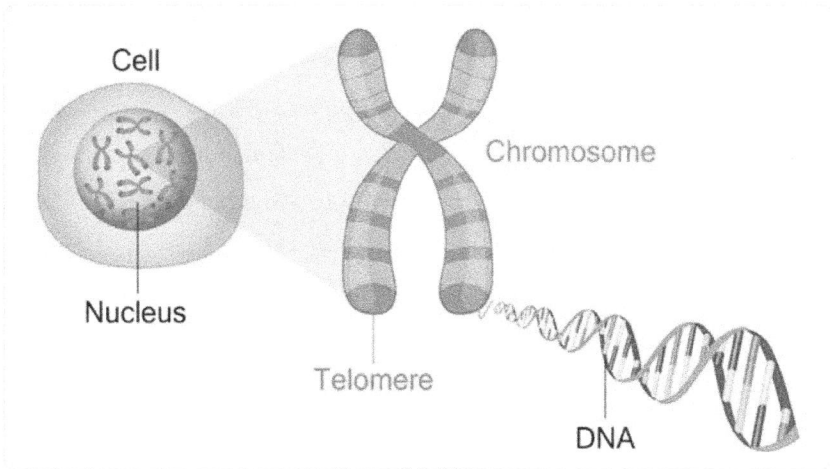

Figure 9.8: Telomere

Source: Author

So, in theory, does that mean that adding telomere to chromosomes can keep cells dividing forever?

Well, cancer cells are already doing that. They sometimes create their own telomerase.[54] They can replicate indefinitely without getting their chromosomes damaged. Because of that, no scientist has the urgency to inject people with telomerase.

Researchers on longevity have focused on researching a type of nematode or roundworm called the Caenorhabditis elegans. This type of nematode is good for study because they only have about 20,000 genes, and their lifespan is only 14 days. Because of this, scientists can easily find out which gene is responsible for aging. [55] In 1993, Cynthia Kenyon, a biologist from the University of California, found out that just one gene makes these nematodes age. It is called DAF-2. [56] So, she experimented with the gene and mutated it so that the gene doesn't work.

Then interesting things began to happen.

Instead of living 14 days, the nematode managed to extend its lifespan up to 28 days. Kenyon also found that another gene, known as DAF-16, had the opposite role. It kept the worm young and healthy by regulating the production of antioxidants, germ fighting proteins and other things that fight off pathogens and stress.[57] Kenyon found that aging gene DAF-2 works by restricting the effect of the longevity gene DAF-16. When you damaged DAF-2, DAF-16 would continue with its business.[58]

But what about human beings?

Many scientists had done a lot of research that focused on one gene that creates growth hormones called IGF-1. When scientists silenced this gene in mice, there was less cell damage caused by oxidation. Organs seemed to be less susceptible to cancer or other age-related diseases. It extended the lifespan of mice by 33%.[59]

Apart from that, since 1940, scientists discovered that a lower caloric intake leads to a longer lifespan. Caloric intakes stimulate IGF-1. In a study published by Nature Communications, a researcher found out that six of the 20 monkeys on a calorie-restricted diet lived beyond 40 years. The average lifespan for laboratory monkeys is 27.[60]

No one is certain exactly why calorie-restriction increases the lifespan of organisms. Some scientists believe that it may have to do with free radicals, released when the body turns food into energy. Free radicals can damage important parts of your cell – DNA. Other scientists believe that think restriction increases longevity by rejuvenating the body's biological clock - a set of genes that change activity to sync with the cycle of day and night.

Immortality

I was born human. But this was an accident of fate – a condition merely of time and place. I believe it's something we have the power to change.

- Kevin Warwick

Humans have been dreaming about cheating death for a millennia. Today, with our rapidly developing technology, we are finding more and more ways to tamper our own biology.

In the past, we made devices, such as prosthetic limbs, spectacles, false teeth and hearing aids, to correct the imperfect parts of our body. In the future, we could use implants to augment our senses or boost our cognitive process with memory chip implants. The ultimate goal of merging humans with machines is to produce a new species of human with vast intelligence, super strength, long lifespan – immortality.

But can we really use technology to upgrade humans?

In August 1998, Cybernetic expert Kevin Warwick, of Coventry University, had a silicon chip implanted in his arm for a computer to monitor him as he moved through the halls and office of the Department of Defense at the University of Reading in London.[61] The implant was a near field communicator that communicates via radio waves with a network of antennas. The aim of this implantation is to determine whether the information could be transmitted to and from an implant. It was a success. Warwick had several electronic devices implanted into his body. One of them allowed him to experience ultrasonic sound input to have the sense of bats. Another one was to interface his nervous system with his computer so he could control a robot hand in a lab in England but experience it in New York.[62]

Transhumanists envision that one day memory chips and neural pathways will be embedded into people's brains. This will help humans bypass using external devices like computers and phones to access data and make complicated calculations. When this happens, the link between humanity and machines will become increasingly blurred.

No one wants to live forever at the age of 95. But if you can rejuvenate the body to 29 or 30, you might want to do that. This can be done in several ways, from genetic engineering to reversing the aging of cells, replacing vital organs with new parts to upload our consciousness to a computer.

Imagine a future when no one dies, and our minds are uploaded to a digital world. They continue to live in a simulated environment with an avatar body but still contribute to our biological world. This is the type of future we might be heading.

But what would it actually take to scan a person's brain and upload their mind?

The main challenge is scanning the brain in enough detail to capture the mind so that we can simulate the detail artificially. As discussed earlier, the human brain has 86 billion neurons, connected by at least one hundred trillion synapses. All of the neurons and their connection together are called connectome. There are hundreds, and possibly thousands, of connections or synapses. Each of them functions in a slightly different way. Some work faster. Some work slower. Some might be more stable over time relatively. Some might shrink and grow during learning. Some neurons spray out neurotransmitters that affect other neurons. In order to copy a mind, all these kinds of interactions need to be mapped.

The science community around the world never took mind mapping seriously until one day in 2014, when neuroscientists scanned the brain of a roundworm called C elegans and actually uploaded its mind into a robot. To our surprise, without any instruction being programmed into the robot, the C. elegans virtual brain controlled and moved the Lego robot. This is the OpenWorm project. The goal is to replicate C. elegans as a virtual organism. [63]

But the entire nervous systems of C elegans is about 300 neurons with 7,000 synapses.

Assume the openworm reaches a point where humans can scan a brain and replicate the mind, how do we scan it?

Currently, we have the best non-invasive scanning method, MRI. To detect a synapse, we need to scan at a resolution of about one micron (i.e., 1 μm). To distinguish each type of synapse, even better resolution is needed. However, MRI relies on powerful magnetic fields. The problem is scanning the resolution required to determine the level of synapse required will probably cook our brain. That is why a new type of scanning technology is required in the future to make this a reality.[64]

The next challenge will be to recreate that information digitally. With the computing power and storage space, we are much closer to achieving this technological capacity. Artificial neural networks are already running on Internet search engines, self-driving cars and smartphones. Although no one has built an artificial network with 86 billion neurons, with the exponential growth in technology growing exponentially, I am not surprised that this will be happening in the 21st century.

Chapter 10
How to Make a Monster

In the original Jurassic Park movie, the mathematician Ian Malcolm predicted that the small changes in complex systems could have big, unpredictable effects. In the movie, InGen created the dinosaurs through cloning - the process used preserved DNA from mosquitoes in amber, and DNA from frogs to fill in the missing genomes. While the idea of creating a wildlife park of de-extinct dinosaurs might be a science fiction concept, and you don't get blood preserved inside mosquitos in amber, the idea of synthetic DNA doesn't end here.

The World's First Self-Replicating Organism

The first self-replicating species we've had on the planet whose parent is a computer.

-J. Craig Venter

In May 2010, the world's leading genomic scientists, J. Craig Venter and his team of scientists announced that he had created the world's first self-replicating organism. The team began with a small piece of laboratory-made DNA, then used a new technique to join them together into the largest piece of synthetic DNA – a loop of one million units in length. The loop of DNA was designed to replicate the genetic sequence of a species of bacterium. To test the DNA, the team of scientists inserted it into a different species of bacterium - Mycoplasma capricolum. The result is that the synthetic DNA is accurate enough to take over the bacterial cell. The "synthetic cell" replicates itself to form a bacterial colony. Venter called this synthetic bacteria *mycoplasma mycoides jcvi-syn1.0.*[1]

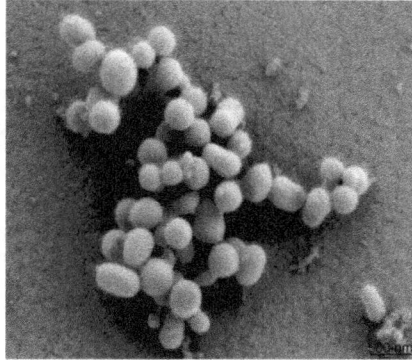

Figure 10.1: Scanning electron micrograph of the world's first self-replicating organism

Source: Tom Deerinck/ Mark Ellisman of the National Center for Microscopy and

Imaging Research at the University of California at San Diego

But is this life? What is the origin of life? How are living things created? How did a single cell bacteria evolve to become a complex organism like us today?

Darwin's Doubt

For over 150 years after the publication of *On the Origin of Species* by Charles Darwin in 1859, Darwinism followers believed that natural selection drives evolution.[2] Darwin's tree of life depicts all forms of life descended from one common primordial form. This primordial branches into other life forms over time, eventually reproducing.

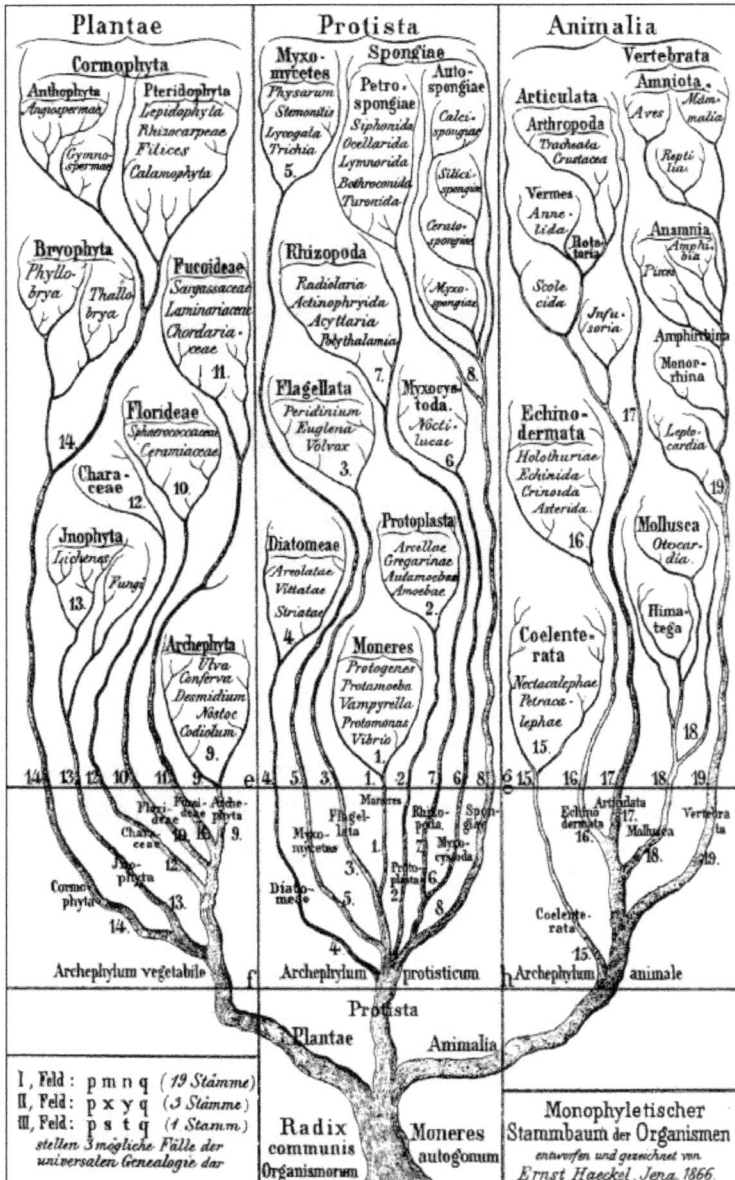

Figure 10.2: Darwin's Tree of Life

Source: peabody.yale.edu

During the long course of ages and under varying conditions of life, organic beings vary in several parts of their organisation, and I think this cannot be disputed. If there is a severe struggle for life, owing to the high geometrical powers of increase of each species, at some age, season, or year, and this certainly cannot be disputed; then considering the infinite complexity of the relations of all organic beings to each other and their conditions of existence, causing an infinite diversity in structure, constitution, and habits to be advantageous to them. I think it would be a most extraordinary fact if no variation ever occurred useful to each being's own welfare, in the same way as so many variations have occurred useful to man. But if variations useful to any organic being do occur, then unquestionably, individuals thus characterised will have the best chance of being preserved in the struggle for life, and from the strong principle of inheritance, they will produce offspring similarly characterised. This principle of preservation, I have called, for the sake of brevity, Natural Selection.
—Darwin summarised natural selection in the fourth chapter of *On the Origin of Species*.

Darwin formulated a theory that explains the branching tree of life – a process called natural selection. Natural selection allows organisms with favorable traits more likely to produce. In doing so, they pass their traits on to the next generation. Over time, natural selection allows organisms to adapt to their environment.

However, when Charles Darwin finished his famous book, he thought he had explained every clue of life, but one: Was life designed or does it merely appear designed?

Can we design life?

DNA is like a computer program, but, far, far more advanced than any software ever created.

-Bill Gates

In 1953, American biologist James Watson and English physicist Francis Crick discovered the double helix structure of the DNA molecule.[3] They also discovered that DNA stores information through a sequence of four nucleotides bases: adenine (A), thymine (T), guanine (G) and cytosine (C).

Uncannily, these sequences store and transmit precise biological information much like machine codes.

Consider the following sequence of letters:

ATGAGCAAG TTCCGAACAA GGATTCGG GGAGGATAGA
TCAGCGCCCG AGAGGGGTGA GTCGGTAAAG AGCATTGGAA
CGTCGGAGAT ACAACTCCCA

Does it looks like a block of encoded information to you?

These strings of characters are not just some random string containing four letters A, T, G, C. They are sequences of genetic assembly instruction for building a protein by living cells.

Now consider another string of characters.

01001000 01100101 01101100 01101100 01101111 00100000 01010010
01100101 01100001 01100100 01100101 01110010 00100001

These 1 and 0 might not look like much to you, but in binary code, the numbers are actually saying "Hello Reader!", written in the binary conversion of the American Standard Code for Information Interchange (ASCII).

Since Watson and Crick discovered the information-bearing properties of DNA, scientists around the world have become increasingly aware that Darwin's theory of natural selection has not yet adequately explained the origin of life. What are inside cells is information in nanoscale. In computer science, if you want to write a function to do something, you need to code. This is the same in life. If you want to create a living organism, you need to give cells information to make it happen. Somehow, it makes anyone wonder if the origin of life is actually by design.

Although Watson and Crick's discovery had solved one secret of how cells store and transmit hereditary information, they led to another mystery about the origin of the information this is required to build the first living organism.

What is this information? Where did life actually come from? What is actually encoded into DNA?

It looks like the question of the origin of life is essentially asking the same question of the origin of biological information.

The Book of Life

In the last 50 years, scientists have come to understand how information in the cell is stored, transferred, edited, and used to construct molecular machines and circuits made of proteins.

Because information inside living things makes life more mysterious, scientists began to map, character by character, and complete the sequence of genetic instruction inside the human genome and other species. In 2000, U.S. president Clinton announced on the White House lawn the completion of the Human Genome Project. Francis Collins, the director of this project, called the genome the *Book of life*.[4]

How did the *Book of life* translate to creating life?

The Mystery of Life

Approximately 541 millions years ago, the Cambrian explosion happened.[5] This was an important time on Earth because it was the time most of the major groups of animals first appeared in fossil record. It was a time when a diversity of life appeared.

What makes a new proliferation of life possible?

Consider a group of single cell eukaryotic or prokaryote organisms swimming with flagellum. What separates it from trilobite with complex characteristics?

And you are right. They are different by the amount of cells. A trilobite has relatively more diverse functions because it has more cell types. Since cells require proteins to survive, these new cell types will require new proteins, as well as instructions to make these new proteins. New proteins that are capable of performing a new function in an organism require new proteins folds. For a Cambrian explosion to happen, a mechanism must have happened that produces distinct new proteins and protein folds.

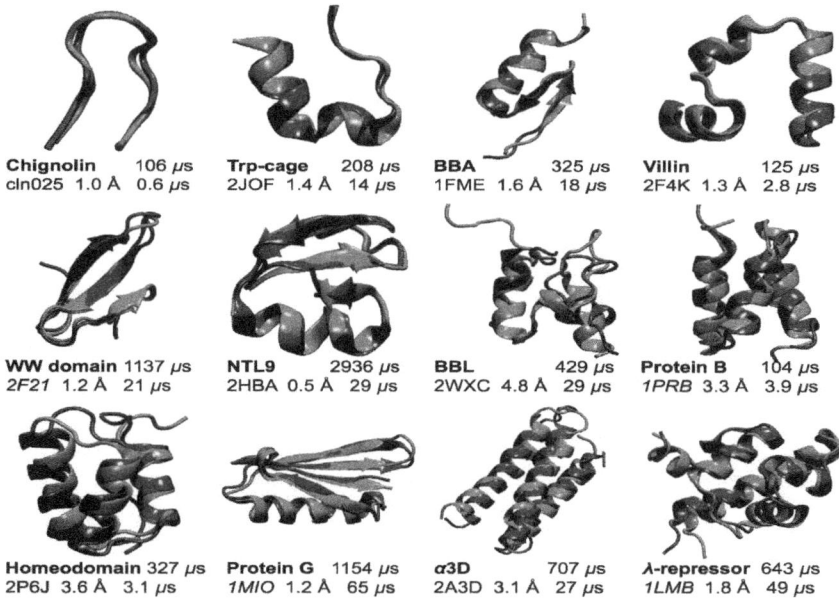

Chignolin 106 μs	**Trp-cage** 208 μs	**BBA** 325 μs	**Villin** 125 μs
cln025 1.0 Å 0.6 μs	2JOF 1.4 Å 14 μs	1FME 1.6 Å 18 μs	2F4K 1.3 Å 2.8 μs
WW domain 1137 μs	**NTL9** 2936 μs	**BBL** 429 μs	**Protein B** 104 μs
2F21 1.2 Å 21 μs	2HBA 0.5 Å 29 μs	2WXC 4.8 Å 29 μs	1PRB 3.3 Å 3.9 μs
Homeodomain 327 μs	**Protein G** 1154 μs	**α3D** 707 μs	**λ-repressor** 643 μs
2P6J 3.6 Å 3.1 μs	1MIO 1.2 Å 65 μs	2A3D 3.1 Å 27 μs	1LMB 1.8 Å 49 μs

Figure 10.3: Protein Folds

Source: science.sciencemag.org

[Note: The instruction to make protein is encoded in our cell's DNA. It can be broken down into two steps: transcription and translation. When the cell needs to make a protein, a copy of the cell's DNA is used as a template called mRNA (messenger RNA), which is created in the nucleus. This process of creating mRNA is known as transcription. The mRNA is then transported out of the nucleus to the ribosomes in the cytoplasm. Translation happens in the ribosomes, which consist of rRNA (Ribosomal RNA) and proteins. In translation, the instructions in mRNA are read, and tRNA (Transfer RNA) brings the correct sequence of amino acids to the ribosome. Then rRNA helps bonds form between the amino acids, producing a polypeptide chain. After a polypeptide chain is synthesized, it may undergo additional processing to form the finished protein. This is protein synthesis.]

You may realize that the Cambrian explosion doesn't just mean an explosion of life. It also means a sudden explosion of genetic information happened during this period. So, if Darwin's evolution theory is correct, what does it takes for natural selection, acting on a random mutation, to build an organism like a trilobite? How do these mechanisms generate the required functional genetic information?

The truth is that to create life, natural selection and mutation must generate a precise array of nucleotide bases to build novel protein structures. Modern genetics had established several techniques like duplications, insertion, inversions, recombination and deletion of genetic text. However, random change to genetic codes that simulate mutation adversely degrades the function of the genetic sequence. It's as if you randomly change a few digital characters in a computer program, most likely, the modified program will halt. Also, assume that random mutation does works, the chance of finding a correct combination nucleotide is prohibitively small. Typically, for DNA to encode most proteins, about 1000 base pairs would be sufficient. But there are extra or nonsense sequences inside genes, which makes the gene longer than necessary. Human genes are commonly around 27,000 base pairs long, and some are up to 2 million base pairs. The smallest genomes, belonging to primitive, single-celled organisms, contain just over half a million base pairs of DNA. Scientists found that there is approximately 1077 possible sequence to specific a functional sequence of 150 amino acids. However, given 1040 new organisms have lived on Earth since life began, it makes anyone doubt if Darwinism's theory of evolution really explains the origin of life. The gene evolution of 1 chance in 1077 over the time of 3.8 billion years to produce 1040 new organisms just doesn't make sense. Darwin cannot explain large-scale macroevolutionary change, as we see in the Cambrian explosion. Micro-evolution through natural selection and mutation change cannot turn reptiles into mammals or convert fish into amphibians. There is a missing link here.

The De-extinction Project

Given the attention for the world's first mammal clone in 1997, Dolly the sheep captured the imagination that we could reproduce the entire animal, if we could just obtain the single working nucleus from any cell.

Figure 10.4: Dolly the Sheep in National Museum of Scotland

Source: National Museum of Scotland

Somatic cell nuclear transfer (SCNT) is a technique used to created clones. Two cells are needed– an egg cell and a somatic cell. During the cloning process, the nucleus of the egg cell is removed, leaving it enucleate. The nucleus of a somatic body cell is then removed in fuse with the denucleated egg cell. After being inserted into the oocyte, the somatic cell nucleus is then reprogrammed by its host egg cell. The ovum, now containing the somatic cell's nucleus, is stimulated with a shock and will begin to divide. The egg is now capable of producing an organism with the necessary genetic information with just one of its parents.[6]

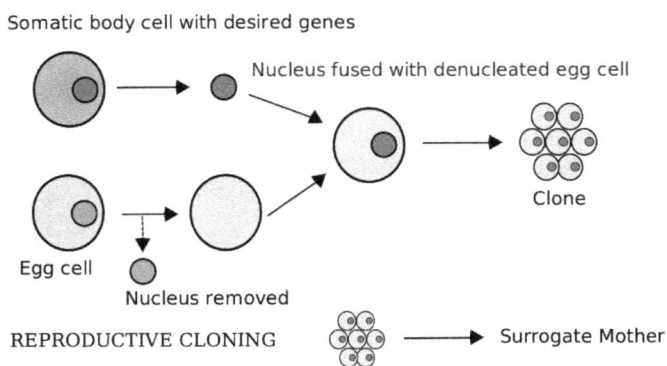

Figure 10.5: Somatic cell nuclear transfer

Source: Author

As remarkable as it sounds, cloning is not always perfect. Dolly is just one clone to be born live of out 277 cloned embryos.[7] Also, there are other side effects like premature aging, the problem with the immune systems, defects in vital organs like liver, brain and heart. Clones created from a cell taken from an adult might already have shorter chromosomes; thus, a shorter life span. That was why Dolly only lived up to six years old, whereas other naturally born sheep lived up to twelve years old.[8]

Although SCNT is famously used in Dolly, the Sheep, the application of cloning has much larger implications. One of them is the de-extinction project.[9] Interspecies nuclear transfer (iSNT) is a technique to rescue endangered species or de-extinct species. Instead of using donor and recipient cells of the same species, cells from two different organisms of closely related species within the same genus are used. Scientists want to find out the possibilities of producing transgenic cloned embryos by iSCNT of cattle, mice, and chicken donor cells into enucleated pig oocytes. The result is that pig oocytles are reported to reprogram the donor cell nuclei from cattle, but they fail in reprogramming the somatic cell nuclei from mice. Pig oocytes could reprogram bone marrow cells from chickens to produce transgenic chicken iSCNT successfully, but chicken cells did not fuse well with the enucleated pig oocytes. Several species are known to show normal chromatin remodeling and embryonic development beyond the embryonic activation stage (EGA), but they fail to develop further due to gene expression.[10]

Despite these failures, scientists in Spain demonstrated the possibility of cloning extinct animals. For example, Bucardo was a subspecies of ibex. It was able to adapt to the extreme cold and snow of winter in the Pyrenees.[11] However, its population had been declining due to hunting and other reasons. In April 1999, researchers captured the last of its kind, a female called Celia. The following year, Celia was killed in an accident. So, a team of scientists injected nuclei from Celia's preserved cells into enucleated goat eggs. They implanted eggs into hybrids between the Spanish ibex and domestic goats. Of the 57 implantations, one was carried to term. It was a baby bucardo, born in 2003, but then died a few minutes later because of a lung defect.[12]

Cloning endangered species using iSCNT is a daunting task. The iSCNT-cloned embryos usually had extremely poor development. In fact, scientists had been doing many trials of iSCNT in wildlife species like the giant panda, Siberian tiger and Sei whale. And the best result of iSCNT in mammals occurred when using sibling species that can hybridize naturally like the river buffalo.[13]

Although bringing dinosaurs back to life using iSCNT might remain science fiction, for now, are scientists on the verge of resurrecting other prehistoric species like the Woolly Mammoths?

In 2010, scientists discovered a mummified woolly mammoth in the frozen permafrost in northern Siberia. Her name is Yuka, one of the best-preserved mammoths in the scientific world. She last lived 28,000 years ago. And it turns out its icy tomb might not be the end of Yuka's story. Japanese scientists extracted Yuka's bone mallow and tissues extracted 88 cell nuclei and implanted them into mice egg cells. It turns out that a number of reconstructed oocytes had signs of cellular activity.[14] However, no further cleavage was observed, possibly because of the damage in nuclei transfer.

The Harvard Wooly Mammoth Revival team led by genetic professor George Church is also working on bringing back a version of the woolly mammoth by combining its DNA with the living Asian elephant.[15] They want to equip the Asian elephant with genetic tools to survive the Arctic tundra. His team had successfully read and reconstructed the complete sequence of the ancient genome of the Wooly Mammoth. They have identified the mammoth for extra fat, dense hair and improved oxygen-carrying capabilities in the blood to survive the frozen north. Church believes that reintroducing this creature might fix the Arctic tundra and stop the release of greenhouse gas emissions from the ground. Mammoth-like creatures trampled mosses and shrubs, uprooted trees, and inadvertently maintained a landscape full of grasses, herbs, and no trees. Because grasslands have been shown to reflect more sunlight than the surrounding larch forest, it would reduce the heat penetrating the ground.

[Note: Arctic lands are covered by areas of ground known as permafrost that have been frozen since the Pleistocene. It contains a vast amount of carbon, 1600 gigaton of carbon from dead plant life locked away by the extremely cold temperature. The amount of carbon in these frozen stores is estimated to be about twice as much as that currently in the atmosphere. They are likened to "sleeping giants' in our climate system.]

Also, ancient DNA holds secrets that impact modern biology. These prehistoric species have genomes that adapt to the years of catastrophes, epidemics and environment changes. Understanding the information locked inside these genomes can possibly be useful for treating human disease, and potentially, the future human space exploration in surviving in cold environments.

An Incomplete History of Genetic Engineering

Imagine you are back in the 1980s and were told that computers would soon take over everything from shopping to dating, the stock market, and billions of people would be able to connect to the web, and you would own a handheld device as powerful as a computer? It would seem absurd. But then it happened.

Science fiction became our reality, without us even knowing about it.

In genetic engineering, we are at a similar point today.

Humans have been trying to engineer life for thousands of years. Through selective breeding, we have strengthened the useful traits of livestock and plants. Until we discovered DNA, we did not truly know how it works. We just know information about life is encoded in DNA. We know that nucleotides are paired and make up a code that carries out instructions. Changing this instruction changes the organism.

In 1960, scientists tried to experiment with random mutation in the genetic codes of plants with radiation. The goal was to get a useful plant variation by pure chance.

In the 1970s, scientists began to insert DNA snippets into bacteria, plants and animals to study and modify them. The earliest genetically modified animal was born in 1974, making mice a standard tool for research to save millions of lives.[16] In 1980, Indian American scientists,

Ananda Mohan Chakrabarty, engineered the first microbe, multi-plasmid hydrocarbon-degrading pseudomonas, in the world to absorb oil.[17]
Today, we produce many chemicals by means of engineering life, such as insulin.

The first genetically modified food in the lab, the Flavr Savr tomato, went for sale in 1994.[18] It was given a much longer shelf life where an extra gene suppressed the build-up of a rotting tomato enzyme.

Today, with genetic modification, we have super muscle pigs, fast growing salmon, see-through frogs, and glow in the dark fish. Fluorescent zebrafish are available for as little as ten dollars.

Even so, gene editing is not cheap. It is very expensive, complicated and time-consuming.

But with CRISPR, this has now changed everything.

CRISPR is a gift from Mother Nature. It is a revolution. In the last chapter, we discussed how CRISPR works and how it is being used to fight viral infection. In reality, this precise gene-editing tool is rapidly transforming genetic research as well. CRISPR causes the cost of genetic engineering to plunge by 99%. Instead of taking years, CRISPR takes a few weeks. With CRISPR introduced to the world in 2012, anyone with lab access can now do genetic experiments unimaginable in history.

A second look at CRISPR

Malaria is a serious-life threatening disease that kills more than 445,000 every year. It infects a certain type of mosquito that feeds on humans.

Over the years, scientists have been trying to solve this problem, but the problem has been harder than expected. Eventually, a US biologist, Anthony A. James, found a way to genetically modify a mosquito by adding a gene that makes it impossible for the malaria parasite to survive inside the host, thus preventing the parasite transmission.[19]

But here is the problem: How do you replace all the malaria-carrying mosquitos?

One way is to breed them, release them into the wild, and hopefully, they can pass on their genes. But for this method to work, you have to have at least ten times the native population of mosquitos to work. So, a village that has 100,000 malaria-carrying mosquitos will need 1,000,000 malaria resistant mosquitos.

That is where CRISPR comes into play. CRISPR could not only guarantee a particular genetic trait would be inherited, it also promises that this trait would spread very quickly. James did an experiment by engineering two mosquitos with an anti-malaria gene (red eyes) using CRISPR and put them into a box with 30 ordinary mosquitos (white eyes) to breed them.

In two generations, there were 3,800 mosquitoes. Given 30 white eyes mosquitos, anyone would have guessed the majority of the offspring would have white eyes. But the surprising fact is that it is not. When James opened the box, all 3,800 mosquitos had red eyes!

If you remember our biology, this shouldn't happen. When a male and female mates, the baby inherits DNA from each parent. So, if one mosquito has the aa gene, and the other one has the aB gene, where a is the normal gene and B is the anti-malaria gene, the offspring will be in these four permutations: aa, aB, aa, Ba.

However, with the CRISPR gene drive, all of the offspring had aB. The gene drive created guarantees a trait would get passed on. What is exciting and frightening about this technology is that the CRISPR gene drive will spread the change in the gene relentlessly until it is in every single individual in the population. In terms of controlling Malaria, this is remarkable. Just a small population of mosquito with the anti-malaria gene released into the population and Malaria will become a disease of the past. It will change an entire species.

[Note: Gene drive is a genetic engineering technology that can alter—and potentially eliminate—entire species.]

So, you see, CRISPR could potentially be used to eliminate invasive species. All that's needed is to produce a gene drive with only male offspring. And in a few generations, there will be no females left.

However, since the CRISPR gene drive is so effective, an accidental release could change the entire species very quickly. Interbred of species with gene drive will have unimaginable consequences to the ecology. And if the CRISPR gene drive is somehow misused, and genetically modified organisms (GMO) are accidentally released into the wild, the consequences could be disastrous. Anyone can use CRISPR. A talented undergraduate with some lab equipment can do this.

Despite these factors, CRISPR technology can also be used for many innovative application, such as creating pets with custom color and size, making allergy-free foods, increasing the production of biofuel by algae by creating strains with higher fat output, improving muscle strength, enhancing the omega-3 content of salmon, or even de-extinction.

How to make a Monster?

So far, we've talked about creating life using SCNT for cloning and CRISPR to edit DNA. Both these techniques borrow too much from an existing organism. They do work, but it doesn't answer the deeper question as to what life actually is. Is life just normal matter organized in a particular way? After all, non-living things and living things are all made up of matter composed of molecules of compounds or elements.

I believe the goal of creating life de novo is actually one of the most exciting things that will happen in the 21st century. Instead of taking a gene from one organism and putting it into another, synthetic biologists are working to build the ideal organism from scratch.

Synthetic biology involves the creation of an entirely new species. Xenobiology is the creation of organisms made from alternative forms of DNA, known as XNA.[20] At the beginning of this chapter, we talked about how scientists had created the world's first self-replicating organism. In fact, scientists can now design artificial life in the same fashion that Frankenstein created his monster, but at a genetic level. They can now design novel living systems and new species from a set of standardized genetic blocks, known as Biobricks.[21] Think of it as genetic Lego, where you can assemble anything like custom chromosomes. These custom chromosomes contain artificial DNAs that can be put into empty cells to create new microbial life that does what we program them to do. In a paper by the Weizmann Institute in Israel, it says there is roughly the same number of bacteria in our body as human cells, an untapped source of potential for reprogramming these bacteria into microscopic medical drones to fight diseases. We can also create synthetic bacteria that can synthesize biofuels to extend its shelf life.

Dr. Craig Venter, the father of synthetic biology, spent 15 years sequencing the DNA of the bacteria and then artificially reproduced it. He reproduced Mycoplasma genitalium (M. genitalium), the smallest bacterium ever existed. Mycoplasma has the fewest gene of any living thing with only 1,097,000 DNA in length, or 525 genes.[22] That makes it the ideal candidate to help scientists understand which functional gene is essential for survival. In 2010, scientists at the J. Craig Venter Institute (JCVI) created an entire organism from scratch. They did this by sequencing the genome of M. genitalium, coding it into computer form, and reconstructing the entire genome with a DNA printer, along with a few biomarkers in the genetic code, so they could track and identify the creature. Once they printed the genome, they injected it into an empty cell to create the world's first synthetic bacteria – Synthia (Syn 1.0). Synthia meets the scientific definition of life. It is able to survive, making energy to grow and reproduce.

Even so, Synthia is not truly synthetic, as the genome is from M. genitalium. So, Dr. Craig Venter had a second attempt. He deleted all M. genitalium's genes and only inserted the ones he considered ideal. The idea is to design a creature with the fewest genes possible – Mycoides JCV-syn 3.0, with only 473 genes.[23]

This experiment is not just about creating the simplest lifeform. With a true understanding of what these genes do, future scientists will be able to help us understand our evolutionary origins, eliminate diseases, immunize our crops growth and design better medical treatments.

Xenobiology

Xenobiology is a sub-field in synthetic biology that describes how scientists are not just creating lifeforms from scratch but creating a new type of life altogether. It involves synthesizing useful biological systems and alien organisms that do not exist in nature. This involves creating new nucleotide letters, substituting new amino acids and even encoding our genes with unnatural sequences of DNA. This diversity of our genetic code will lead to an invention of new, novel proteins and create new possibilities for life.

Synthetic biologists do this by creating brand new alien organisms out of molecules not based on DNA but on XNA (Xeno nucleic acids).[24] If you remember in the last chapter, all life forms are governed by genetic code with just four letters: A, C, G and T. The DNA of all life on this planet only codes for 21 amino acids and assembles them into protein using a translational machine of our cells called ribosomes. Every three letters of our RNA, such as "GCU" or "ACG," make up a codon. There are 64 possible codons because there are only 64 possible ways to rearrange A, C, G and T. The ribosome than pairs each of these 64 codons with specific amino acids to create a protein chain. Once a specific amino acid chain is made from our gene, our cells will then fold it into a unique protein found in our body. However, there are only 21 amino acids. So, multiple codons are forced to share the same amino acids.

Dr. Floyd Romesbery, an American biologist, believes that rewriting the unused genetic codons to correspond with new amino acids will create new proteins for medical applications. This can only be done if we pair preexisting codons with new amino acids.[25]

Perhaps the most exciting thing about xenobiology is the idea of adding new DNA letters to diversify our genetic codes. In 2015, the American

Chemical Society announced the creation of two new letters, P and Z.[26] A paper published in the journal Nature, led by Dr. Floyd Romesbery at Scripps Research, announced they had successfully introduced two artificial nucleotides into bacteria. These new bases are called X and Y, which sit well with the bacteria's natural DNA.[27]

"In principle, we could encode new proteins made from unnatural amino acids, which would give us greater power than ever to tailor protein therapeutics and diagnostics to have desired functions."

-Floyd Romesburg

So, by adding new genetic code with just letters A, C, G and T, scientists can use XNA rather than DNA to carry out genetic information. These molecules would have the potential to reveal our knowledge of biology and the origin of life, incorporate into medicine as disease-fighting agents and coat into GMO crops to make them immune to virus.[28]

But XNA cannot transcribe by ribosome since our body can only recognize the four natural nucleotides. To do so, synthetic ribosome must be created. In the next stage, through building synthetic ribosomes to reassign genetic codons to produce brand new amino acids, we can create new artificial proteins. These artificial proteins, designed on a computer and synthesized in the lab, could be used to build brand-new biological circuits inside living cells and transform it into smart cells with a novel, marvel ability.[29]

Right now, the most common XNA are PNA (Peptide nucleic acid), Glycol nucleic acid (GNA) and TNA (Threose nucleic acid).[30] It has been hypothesized that the earliest life form may have used PNA as genetic material as opposed to RNA or DNA due to its extreme robust, simpler formation.[31] If this is true, our general knowledge of RNA/DNA as the earliest life form might need to be rewritten.

More ways to Make Monsters...

In Greek mythology, the term chimera refers to a two-headed creature that contains the appendage of both a goat and a lion. Although chimeras might look creepy in fantasy, in reality, they actually exist.

In science, a chimera is made by merging two different zygotes. If one of those zygotes is human and the other is from an animal, it could help us to grow body parts inside farm animals. Chimeras are defined as organisms that contain the genetic code of more than one individual by fusing cells from different species. For example, if you merge a sheep zygote with a goat zygote in the very early stage of embryonic development, you will create a chimera that exhibits the traits of both a sheep and a goat.

Chimeras can be used to grow human hearts inside pigs. These farm animals are used as bioreactors to grow genetically identical organs for ourselves. This is done by injecting human stem cells into pig embryos and then implanting these embryos into female pigs. At the end of these experiments, scientists had successfully developed human tissues in pig organs. However, a pig fetus takes 16 weeks to develop while a human fetus takes 40 weeks. Growing functioning human organs in pigs resulted in premature failure.

Splices are interspecies hybrids that are artificially modified with humans or genes with other animals.[32] Interspecies hybrids do exist in nature. Liger is a hybrid of lion and tiger. It has the stripes of a tiger, but also keeps the lion mane. Jaglion is another hybrid of jaguar and lion. Mule is the hybrid between a donkey crossed with a horse. These creatures are interesting.

To take it one step further, biotechnology allows scientists to slice individual genes from different animals to create more custom splices.

The most popular splice today is probably the Spidergoat - a goat genetically engineered to make spider proteins in its milk. The spidergoat was invented by Dr. Randy Lewis from Utah State University, looking for a way to farm very special spiderwebs called dragline silk.[33] This is a marvelous material that is one hundred times stronger than human ligament but light as a feather. Spidergoat is created by taking six different genes from spiders responsible for creating dragline silk and splicing them into fertilized goat eggs. These silks could one day be used in a military bulletproof vest or the strongest industrial compounds.

Should these spider genes splice into human DNA, maybe one day, we can make these proteins too. But don't get too excited because I don't think we can become Spiderman who can shoot his webs.

Mythical creatures like the Minotaur, Unicorn and Centaur might be legends, but hybrid engineering has now opened a pandora's box to a world of possibilities.

Twenty years into the 21st century, scientists are already engineering hybrid creatures that don't exist in nature. In the future, we might see splice zoos with a Griffin, Pegasus and even Mermaids. But let's hope that it doesn't end the same way as in Jurassic Park.

Chapter 11
Nanotechnology

Nanotechnology is going to be a major driving force in the technological revolution in the 21st century. When combined with AI technology, we are now at a threshold where we can manipulate matters at an atomic level. This will have tremendous impact on our society, our economy, our environment, our health system and almost all sectors in the manufacturing industry.

How small is a Nanometer?

We live in a macro world, where the concept of a nanometer seems abstract. To put it into perspective, an ant is about 5 mm. The diameter of your hair is about 60 microns (μm). If you scale down, our blood cells are about 2-5 microns. A nanometer is a thousand times smaller than that. It is invisible to our naked eyes. Virus (100nm) and DNA (2nm) are beings in the nanoworld.

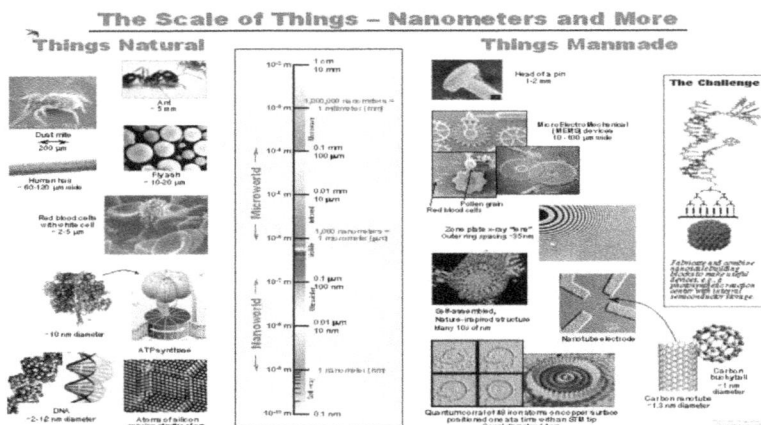

(From NNI website, courtesy Office of Basic Energy Sciences, U.S. Department of Energy.)

Figure 11.1: The scale of things

Source: NNI, courtesy Office of Basic Energy Sciences, U.S. Department of Energy

But, why make things so tiny?

Smaller is Lighter, Faster, Smarter

If you still remember from previous chapters, a computer in 1940 filled the whole room. It is bulky and does not have too much computation power. Today, we have laptop computers a million times more powerful than that.

Moore's law allows us to shrink the size of transistors. The smaller we shrink the size of transistors to nanoscale, the more transistors we can pack into a microchip, and the more powerful the computer we can create.

Well, nanotechnology is not just about making things smaller for the sake of it.

The main fundamental reason why nanotechnology is so different is that the rule in science in nanoscale is very different from its macroscale counterpart. Being able to manipulate matter at an atomic level means a whole new range of technology will be possible in the 21st century.

Science has Different Rules in Nanoscale

Take gravity as an example. Gravity dominates everything on Earth. It is necessary for the rain to fall or water to drain. If you pick up an apple and drop it, it will fall due to gravitational force. But if you pick up a nanoparticle and drop it, will it fall the same way?

On a nanoscale, gravity is negligible. It is weaker than other forces like the electromagnetic force between atoms or molecules. Nanoparticles are more sensitive to force like the Brownian motion or turbulent diffusion. So, it won't fall like apples do.

Another example is that Gold, the precious metal, is a golden color in macroscale, but Gold is not the same color when viewed in nanoscale. Nanogold behaves very differently with light. It can look red, orange or even blue due to the quantum effect.[1]

So, a material can act very differently when it is nano-sized.

If we learn these rules in the realm of the nanoworld, it is possible to make or manipulate them to behave the way we would like it to be.

The Building Block of Nanotechnology

Throughout history, humans have always been trying to make compounds to form new substances to move civilization forward. The Bronze Age marked the first time humans started to combine elements to make an alloy. And in the last century and a half, we have learned how to make synthetic polymers using cellulose or plentiful carbon by petroleum, which we called plastics. Bronze, steel, nylon and plastic are just some of the alloys or synthetic polymer that changed our world today.

In the infancy stage of nanotechnology, we are beginning to enter a new paradigm, where scientists can control matter at an atomic level.

The materials used in nanotechnology are called nanomaterials, which are less than 100nm. They can be carbon-based material in shapes like hollow spheres, cylindrical, rod shape, tubes, etc. They have many potential applications like making stronger but lighter material or electronics.

One such nanomaterial is graphene.

Graphene is an allotrope of carbon discovered by two researchers at the University of Manchester, Professor Andre Geim and Professor Kostya Novoselov, in 2004.[2] It is a one-atom-thick layer of carbon atoms arranged in a hexagonal lattice, which is stronger than diamond. It is the thinnest material on Earth with one atom thick but also incredibly strong. It is a semiconductor, but when it behaves as a conductor, it is an excellent conductor of heat and electricity. It also has interesting light absorption abilities, which makes it transparent. That is why graphene is known as a *wonder material*.

[Note: At room temperature, graphene can conduct electricity 250 times better than silicon. It conducts heat 10 times better than copper, the most commonly used conductor in electronics.]

Figure 11.2: Graphene

Source: Author

A team of researchers in Nature Communication found that graphene-based transistors could actually work better than silicon transistors we use in conventional computers today. This means graphene could be a potential candidate to replace silicon transistors in electric circuits in the future.[3] This could extend the life of Moore's Law.

If we want to continue to push technology forward, we need faster computers to run bigger and better simulations for climate science, space exploration, and Wall Street. To get there, we can't rely on silicon transistors anymore because it is reaching its physical limit. The processing speed of microprocessors using silicon transistors has been stuck at 3-4 GHz since 2005. This is the limit to the rate of signal and power that silicon transistors can handle due to its resistances. But graphene-based transistors could improve these clock speeds up to a staggering 100GHz and requires only one-hundredth of the power required.[4]

Beside transistors, graphene's fundamental properties allow it to be a potential candidate to be used in energy storage, particularly supercapacitors. Nanoarchitectured Graphene-based supercapacitors could enable cars to drive as far as a diesel vehicle and recharge within minutes.[5] A graphene battery could dramatically increase the lifespan of a battery compared to a traditional lithium-ion battery.

What is more interesting with graphene-based nanomaterials is its biocompatibility.[6] It could be used in biomedical applications like implants for brain computer interfaces. If you have a damaged eye, the biocompatibility of graphene transistors could let us create a mesh that interacts with your optic nerve, helping to send image data to the visual cortex.[7]

Potential Unlimited Energy Source

University of Arkansas researchers have demonstrated that the motion of graphene could potentially supply an unlimited source of clean energy. Professor Paul Thibado and his team showed that graphene has a 'rippling effect' in its freestanding form, with each flipping up and down when under ambient temperatures. It works by having a negatively charged sheet of graphene (roughly 10 microns across) suspended between a pair of metal electrodes. When it flips up, it produces a positive charge, and when it flips down, it creates another positive charge, generating an AC current. The movement inverts graphene's curvature and creates free energy.[8] Professor Paul Thibado called this *Vibrational Energy Harvestor* - a machine that harnesses the power of fluctuated graphene and transforms it into electricity.[9]

Imagine the things you could do today if you never had to replace a battery again? Things like smartphones, IoT devices could be powered solely by the heat of room temperature; it would revolutionize everything.

Carbon is the fourth most abundant element in nature, so if we scale up the *Vibrational Energy Harvestor*, it is entirely possible to create a revolutionary decentralized power system across the globe, thus, solving the energy crisis in third world countries.

Solving the Water Crisis

As described in an earlier chapter, one of the biggest problems in the future is the availability of clean water to the increasing population in the 21st century.

By 2050, water consumption will have grown by 40%. A significant portion of the world's population will not have access to clean water. A majority of this population will be in Africa. Even today, millions of children in third world countries die from diseases associated with drinking contaminated water or inadequate freshwater supply. Since 96.5% of water comes from the ocean, to get more freshwater, we would have to pull salt out from the ocean to make it drinkable.

Graphene can help solve this problem.

Graphene is the water filter of the future. Scientists at Australian research centre CSIRO have used graphene to create a simple filtration system that could change the lives of millions in the developing world by making the process of purifying water faster and more effective.[10]

At present, the only solution is an industrial scale of a process called Reverse Osmosis (RO). Reverse Osmosis works by using a high-pressure pump to force water molecules through a semi-permeable membrane. During this process, the contaminants and salts are filtered out, leaving fresh, drinkable water, but the cost of using RO is expensive. This is because this method uses a lot of energy to pump water through the plastic membranes of the cartridge filters. Replacing these plastic membranes with graphene sheets will reduce the energy needed for RO and make freshwater much cheaper.[11]

In fact, researchers at MIT had already built a nanometer scale graphite porous just wide enough for water molecules to pass through.[12] With this solution, an industrial scale water filter plant can use this type of technology and perhaps the filtration system at home too. This revolution graphene filter could solve the world's water crisis.

But the application of graphene doesn't stop here.

Solving Air Pollution

Graphene doesn't only filter water. It has other applications like storage, separation and purification of gas and liquid in industries. Scientists have been attempting to invent a type of filter that graphene-based membranes for carbon dioxide separation.[13] If this succeeds, this novel membrane technology will help to deconstruct carbon dioxide in the atmosphere, capturing radioactive waste particles in our environment.

A Solution to the Fukushima Daiichi Nuclear Disaster

The Fukushima Daiichi Nuclear Disaster was started by the Tōhoku earthquake and tsunami on Friday, 11st March 2011. A large quantity of radioactivity was released into the environment. This radioactivity includes small, poorly soluble cesium-rich microparticles.

Nine years have passed, and although the disaster zone is cleared from immediate danger, the cleanup operation is progressing at a very slow pace. Cleaning up nuclear waste is complicated, expensive, and the cost can easily skyrocket past the cost of building the nuclear plant itself. The radioactive material is dangerous and unstable. Plutonium, a byproduct of uranium, produces gamma rays at extremely dangerous levels to humans, and even a small amount can be fatal.

However, a new discovery, reported in the journal Physical Chemistry Chemical Physics, could be a possible solution to cleaning up this disaster. Scientists discovered that graphene oxide (GO) could rapidly remove radioactivity from contaminated wastewater.[14] GO in flakes binds quickly with man-made radionuclides and condenses them into solids. So, when there is a large pool of radioactive material like Fukushima, adding GO can get back solid material from what were just ions in solutions. Graphene oxide burns very rapidly and leaves a cake of radioactive material you can then reuse.

Besides cleaning up contaminated radioactive wastewater, graphene nanoparticles can also capture uranium in the ocean.[15] Uranium is the main source of nuclear energy. There is around 500 times more uranium in seawater than land-based sources.[16] Extracting uranium from the ocean is far greener than ore mining.

Harvest Xenon for Space Exploration

A graphene filter can help us extract industrial quality xenon to accelerate laser technology and deep space exploration.[17] Xenon is the noble gas. Its inert atomic properties make it less corrosive to the ion engine than cesium fuel. However, xenon is rare in reserve, so just one space mission would use up 10% of the global annual production rate. Fortunately, xenon is naturally occurring in the Earth's atmosphere. If we can design a nano selective filter, with pores within the lattice small enough, we can harvest it passively.

Space Elevator

The idea of building a lift that could travel from Earth into Space may sound like science fiction. However, this idea has been around for more than a century. Nanotubes have the potential to use their high tensile strength to be a candidate for a space elevator.

The high tensile strength of nanotubes holds promise to be used as the material for building a space elevator – a long cable strung between a geostationary satellite and an earth based station, allowing astronauts to halt building material into low earth orbits very inexpensively. So far, no Earth material has been able to create a cable strong enough to survive the tidal force of the Earth's gravity.

A research team from Tsinghua University in Beijing has developed a fibre that is so strong it could be used to build an elevator to space.[18] They said that one cubic cm of fibre made from nanotube would not break under the weight of 160 elephants. However, that tiny cable weighs just 1.6 grams.[19]

Figure 11.3: Space Elevator
Source: www.spaceconnectonline.com.au

Cloaking

Metamaterials are artificially designed subwavelength composites engineered to have extraordinary optical properties not found in nature.[20] The word "meta" is used to indicate the unique characteristic of this material. Rather than hard and strong, these materials are known for their unconventional abilities like preventing waves of electromagnetic radiation from scattering upon contact. In other words, cloaking.

If you have watched the movie Predator or Harry Potter, you will be familiar with how cloaking works. But an invisibility cloak might no longer only exist in science fiction; nanotech could make it a reality.

Metamaterial that can reroute electrical waves around an object is called negative refraction, giving it an optical illusion called cloaking.

Either a body absorbs light, or it reflects or refracts, or does all these things. If it neither reflects nor refracts nor absorbs light, it cannot of itself be visible

- **H.G, Wells, 1881**

Fifty years ago, Russian physicist, Victor visa Lago, predicted that the existence of metamaterials will be an emerging field in nanotechnology. This vision is finally starting to come true. Scientists have now designed an ultra-thin material, 80nm in diameter, with brick-shaped gold antennas that counteract light distortion. Any object behind them is hidden.

However, at present, this technique could only be used to hide microscopic objects. To get a Harry Potter style cloaking, scientists would need to find a way to scale it up for meaningful application.

Chapter 12
The Future of Transport

Mobility has always been a fundamental component in part of the economic and social life of any society. Among other technologies, transport is perhaps one of the largest technological innovations. Our current transport system is the outcome of a long historical evolution by periods of rapid change in transportation technologies driven primarily by global trade.

In the 19th century, a transportation system was formed from the development of steam engine technology. This was further expanded in the 20th century with the setting of global air transport, container shipping, and telecommunication networks. With a rapid reduction in time to travel from one place to another, people are more connected than at any time in history. Transportation is a key to globalization for countries around the world. Larger cargo ships mean more goods can be transported between countries at a cheaper rate. It boosts trade and contributes to the global economy.

With global population growth increasing exponentially, demanding faster and better transportation systems, and, at the same time, the world is facing increasingly problematic traffic congestion, energy and environmental problems. It is interesting to see how the transportation system will evolve in the 21st century.

Driverless Cars

A self-driving car, also known as a driverless car, is a vehicle that uses AI to implement autonomous driving. This new technology uses intelligent path planning technology, computer vision and GPS that takes passengers from one place to another without human intervention.

In 2015, Tesla, the dominant electric carmaker in the industry, invented an autopilot function that allows its model to match speed to traffic conditions, keep in a lane, automatically change lanes, transit from one freeway to another, exit a freeway, self-park, as well as summon it to and from your garage.[1] Tesla's founder Elon Musk, also envisioned a future that self-driving cars would one day make money for their owners by autonomously transporting people when the owners are not using them, much like a driverless Uber. In addition, Tesla expects to launch the first Robotaxi network as part of its broader vision for autonomous ride-sharing.[2] A Tesla vehicle owner can enlist their vehicle to join the Robotaxi fleet when they are not using it.

Besides Tesla, Waymo, formerly the Google self-driving car project that is now a business operated under alphabet, had also launched its robot taxi service.[3] Unlike Tesla's business model, Waymo's Robot taxi service operates as a driverless taxi service.

But a driverless car is not limited to transporting people only.

Neolix Technologies, one of the many self-driving logistics startups based in Beijing, had raised $29 million to mass-produce autonomous shuttles.[4] Amid the COVID-19 lockdown in China, Neolix had deployed its vehicles in ten major cities in China for mobile delivery and disinfection. It helps customers reduce physical contact and address labor shortages during quarantines and travel restrictions.

[Note: The Society of Automotive Engineers (SAE) defines 6 levels of driving automation, ranging from 0 (fully manual) to 5 (fully autonomous). These levels have been adopted by the U.S. Department of Transportation.]

Figure 12.1: Neolix Technologies' self-driving shuttles
Source: Neolix Technologies

Hyperloop

In 2017, Elon Musk, the founder of SpaceX and Tesla, held an event at SpaceX's California headquarter, showcasing the latest technology he thinks will revolutionize the future of transportation – Hyperloop. Hyperloop will be the fifth mode of transport beside rail, road, water and air. It will be faster than a plane, safe, low cost, resistant to earthquakes, sustainably self-powering and is not disruptive along the route.[5]

[Note: The Hyperloop Genesis paper conceived a hyperloop system that would propel passengers along the 350-mile (560 km) route at a speed of 760 mph (1,200 km/h), allowing for travel time of 35 minutes, considerably faster than current rail or air travel times.[6]]

The basic idea of Hyperloop as envisioned by Musk is that the passenger pod travels through a tube, either below or above ground. To reduce friction, most of the air is removed from the tubes by pumps. In Musk's model, the pressure of the air inside the Hyperloop tube is about one-sixth of the pressure of the atmosphere on Mars. This means that an operating pressure of 100 Pa will reduce the drag force of the air by 1000 times, using air ski bearings or Maglev for levitation. An external linear electric motor is used to accelerate the capsule to subsonic velocity. By placing solar panels on top of the tubes, the Hyperloop can generate excess energy, more than it needs to operate. Batteries are used to store energy in case of cloudy weather or night. Also, unlike a ground-based high speed rail system susceptible to earthquakes, the Hyperloop system can dramatically mitigate earthquake risk by building a system on pylons where the tube is not rigidly fixed at any point.[7]

If you think the biggest advantage of a future Hyperloop system is getting to your destination faster, think again.

Hyperloop can travel to and from cities in a matter of minutes. It allows someone to live in L.A. but work in San Francisco. Because people can travel much faster from one place to another, companies can hire a broader pool of talent and provide more business opportunities. Apart from that, a collaboration between Virgin Hyperloop One, a U.S. based company, and Emirati supply chain firm DP World, is working to use Hyperloop for cargo, which can send cargo at a speed of 1000 kmph (621mph). It is predicted that a four-day truck journey could be cut to just 16 hours. Its cost will be 50% higher than the truck, but at least five times cheaper than air freight with fewer delays.[8]

Figure 12.2: Dubai- A testing ground for future transport
Source: DP World Cargospeed

As the population grows, traffic congestion and emission will get worse. Imagine all cars on the road today doubled. Even with lane expansions, highways will eventually get congested. Hyperloop might be a possible solution to clear congestion.

So, next time you are stuck in traffic for long hours, think about all the free time you will gain if you can cut that travel time significantly. If you want to reduce your traffic time, support building a Hyperloop.

Even Hyperloop holds a lot of promise that might ease traffic congestion as the world continues to urbanize. In reality, this futuristic technology might take some time before the world will see its arrival.[9]

BFR

In September 2017, SpaceX CEO Elon Musk revised his plan for an enormous new rocket to go to Mars in 2022.[10] This new rocket is called Big Falcon Rocket (BFR). In the conference, he ended his talk with an incredible promise: using the same interplanetary rocket system for long-distance on Earth. Musk claims that these rockets can shuttle passengers from New York City to Shanghai in 39 minutes and go "anywhere on Earth in under an hour" for about the same price of an economy airline ticket.[11]

[Note: BFR is now renamed as Starship.]

TIME COMPARISONS TO MAJOR CITIES

ROUTE	DISTANCE	COMMERCIAL AIRLINE	STARSHIP
LOS ANGELES TO NEW YORK	3,983km	5 hours, 25 min	25 min
BANGKOK TO DUBAI	4,909km	6 hours, 25 min	27 min
TOKYO TO SINGAPORE	5,350km	7 hours, 10 min	28 min
LONDON TO NEW YORK	5,555km	7 hours, 55 min	29 min
NEW YORK TO PARIS	5,849km	7 hours, 20 min	30 min
SYDNEY TO SINGAPORE	6,288km	8 hours, 20 min	31 min
LOS ANGELES TO LONDON	8,781km	10 hours, 30 min	32 min
LONDON TO HONG KONG	9,648km	11 hours, 50 min	34 min

Figure 12.3: Time Comparison to different cities with BFR from SpaceX
Source: spaceX

The Future of Flying Cars

A flying car used to be a science fiction vehicle that only exists in movies. Today, the future of the flying car we have been dreaming about is closer than you think. In fact, some of the flying car prototypes already exist today.

In 2017, Dubai's autonomous aerial passenger taxi, the volocopter, has taken flight in the UAE. It can fly for 30 minutes at a maximum speed of 100km/h.[12] The German startup Volocopter will come autonomously. Passengers can download the app, order a Volocopter to the next Volopod, then the volocopter will pick the passenger up and drop him or her to the destination. The Volocopter does not run with remote control guidance and is equipped with an emergency parachute. Volocopter saves time. It is extremely safe, emission-free, and quiet. It will redefine urban air mobility in the world.

Figure 12.4: Flying car - Volocopter

Source: Volocopter.com

Apart from Volocopter, Global ridesharing giant Uber also plays a huge role in the vertical take-off and landing (VTOL) industry. A network of distributed Skyport will be built to enable Uber Air's operation to handle up to 1,000 landings per hour. Elevate Cloud Service will be used to manage the dense operation of unmanned, low altitude traffic.[13] Ultimately, Uber's plan is to give the rider the option of an affordable shared flight and alleviate congestion on the ground.

Ironman Suit

Perhaps Marvel's Iron Man might not be just another fictional character on the big screen. On September 2, 2020, a pilot above Los Angeles reported an unidentified man flying on a jetpack. It happened 3,000 feet above Los Angeles International Airport. And that man was about 30 yards away from the aircraft.[14]

Jetpack is a VOTL personal aircraft. It uses jets of gas to propel the wearer through the air. While this concept has been in video games or movies for a long time, the first public flight jetpack was revealed as early as June 8, 1961.[15] It quickly captured the imagination of the public. However, because the propulsion technology wasn't mature to create enough thrust, jet propulsion has never become part of our daily lives. It wasn't until the beginning of the 21st century when jetpacks have become a reality once again.

Founded in 2017, Gravity.co was a company pioneering aeronautical innovation. The founder, Richard Browning, built an Iron Man-like suit that leans on an elegant collaboration of mind, body and technology, bringing science fiction dreams a little closer to reality.[16] The suit was built with a micro gas turbine equipped with 1050 HP, pushed through five kerosene-powered turbine engines – one on the pilot's back and two on each arm. It is an entirely pure form of complete three-dimensional freedom.

Figure 12.5: Jetpack

Source: Gravity.co

Besides Gravity, Jetpack Aviation had developed a similar kind of jetpack. The founder, David Mayman, sees that current battery technology doesn't provide very long flight time. In the future, when battery storage density improves in the decade ahead, Mayman is optimistic that the autonomous version of personal transport like a jetpack will roam the sky. Everyone will need a pilot's license. It could shape the future of personal transport in a revolutionary way.

Chapter 13
Industry 4.0

In times of big growth, there is big manufacturing revolutions. It happens three times for every fifty to sixty years. It happened in the mid-19th century with the invention of the steam engine. The mass production model happened in the early 20th century. In 1970, the first automation revolutionized everything.

These revolutions change the way we work in the world of labor. It generates big growth in our economy.

Industry 1.0

In the mid-19th century, steam and water-powered machines were produced to aid workers. As production capability increased, business also grew from individual cottage owners taking care of their own to organizations with owners, managers and employees serving customers around the world. The first industrial revolution had begun.

Industry 2.0

The second industrial revolution happened at the beginning of the 20th century, when electricity became the primary source of power. It is a lot easier than steam and water because the power source is concentrated.

During this period, we have seen improvements in efficiency in manufacturing facilities. The division of labor and mass production of goods in assembly lines became commonplace. Lean manufacturing principles further refined the way in which manufacturing companies improved their quality and output.

Industry 3.0

In the last few decades of the 20th century, the invention of transistors and IC made automation possible. Software systems like PLC control hardware in machines in different industries like food and beverages, mining and petrochemical, etc. However, the pressure to cut costs caused manufacturers to move the assembly operation to low-cost countries like China. This extended geographic dispersion results in the "Made in China" wave today.

Today's Failing Manufacturing Model

Today, the manufacturing model is reallocating factories offshore to places like China in order to reduce costs and take advantage of the cheap labor. This does not inspire production. It is only saving money for a short period and shipping jobs offshore. And cheap labor won't stay cheap for long.

In the last 50-60 years, our existing manufacturing model has just changed the location, the size and the way of operation. We have done everything we can to fine-tune every aspect of this model of operation.

Now, we have reached the limit.

The Internet has changed how we operate in all aspects of life, like the service sector, the media, entertainment, and the way we communicate. However, in terms of productivity, it didn't do much at all. From 1960 till today, productivity growth has been declining sharply, despite all the innovation effort, because of the existing manufacturing model.

Truth be told, all social media today is making us less and less productive. What could we produce by using Facebook and Instagram, apart from gaining likes?

This has to change.

And that is why we need to reinvent the manufacturing base by integrating manufacturing and technological innovation – industry 4.0.

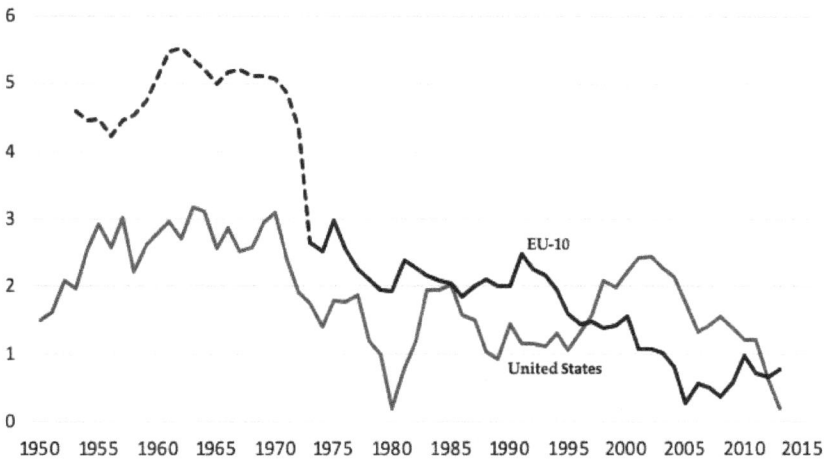

Figure 13.1: Productivity decline from 1960 to 2010

Source: EUKLEMS data except for EU10 1950-72 for the total economy, which is GDP per hour from the Conference Board Total Economy Database, weighted together for the ten EU nations using real GDP weights.

Industry 4.0

Industry 4.0 (I4) first originated in 2011 from a project in the high-tech strategy of the German government.[1] It is the fourth industrial revolution that concerns industry. Let me give you some examples to illustrate how I4 works.

In the aerospace industry, fuel nozzles are some of the most complex parts to manufacture. It requires twenty different parts separately produced and painfully assembled.[2] Now, aerospace companies are using 3D printers that allow them to turn those 20 parts into just one, thus increasing productivity by 40%.

Another thing about Industry 4.0 is that I4 allows more customization and scaling unlike our existing manufacturing model. Imagine a future where we can buy a product design exactly the way we want with the exact functionality we need, and, at the same time, with the same cost and lead-time as it is mass-produced.

In I4, this is possible.

Advanced robotics can be programmed to perform any product configuration without any setup time. 3D printers can instantly produce any customized design. And the production cost and lead-time of one batch of your product is the same as many. The technologies below are just a few of many that have a role to play in Industry 4.0.

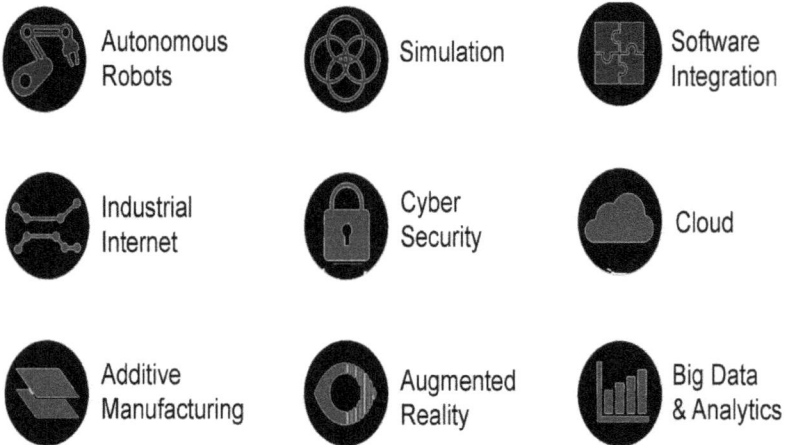

Autonomous Robots   Simulation   Software Integration

Industrial Internet   Cyber Security   Cloud

Additive Manufacturing   Augmented Reality   Big Data & Analytics

Figure 13.2: Industry 4.0

Source: Author

Industry 4.0 factories have machines augmented with wireless connectivity and sensors, connected to a system that can visualize the entire production line and make decisions on its own. With this kind of model, not only manufacturing will become more productive, but it will also become more flexible. I4 factories will be operating on a multi-product made to order basis. Flexibility and customer proximity will drive the future manufacturing model. No longer should China become the factory of the world. No longer should the world produce more than we need to consume. No longer should businesses pile up stocks and make products travel around the world before reaching the customer. The new manufacturing model is decentralized and produces right next to the customer market.

Under the new I4 model, manufacturing businesses will move back to their own country, producing more employment and growth locally.

But, won't Industry 4.0 kill jobs?

Robot production has increased, but at the same time, costs have fallen. Over the last 30 years, the cost of robots has fallen by half in real terms, while labor costs have more than doubled. A lot of repetitive, labor-intensive jobs still active today will be obsolete. Robots and narrow AI will replace many white-collar and blue-collar jobs.

Index of average robot prices and labor compensation in manufacturing in United States, 1990 = 100%

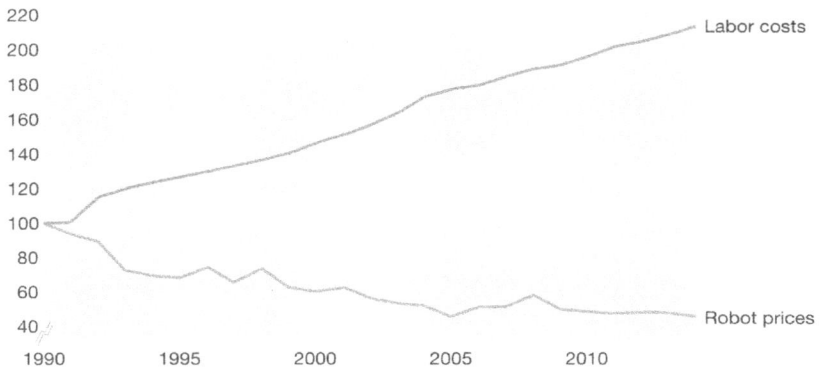

Source: Economist Intelligence Unit; IMB; Institut für Arbeitsmarkt- und Berufsforschung; International Robot Federation; US Social Security data; McKinsey analysis

McKinsey&Company

Figure 13.3: Cost of robot and labor price

Source: Economist Intelligence Unit; IMB; Institut fur Arbeitsmarkt - und Berufsforschung; International Robot Federation; US Social Security data; McKinsey analysis

The future might look dire for workers. Imagine you didn't go to school and learn the right things. It looks like there are less and less jobs among the exponential growth in population, and it is true. Today, your employer looks at your education and your years of experience, but the way how you leverage technology can change that too.

What If even your Grandma can fix a Helicopter?

Imagine if a helicopter is stuck in Africa at a remote place? You had an augmented glass on your head connected to a central computer that would know every detail about that helicopter. The computer will talk to the sensors on the helicopter and tell you where to place the tool and exactly which part to repair through a digital overlay in a hologram. So, you could basically perform the repair action without knowing much about the helicopter. It's as if you are looking through a picture book.

To do so, you don't need a four-year degree in engineering. As long as a person is comfortable with technology, even your Grandma can fix a helicopter if it is the first time she touches one.

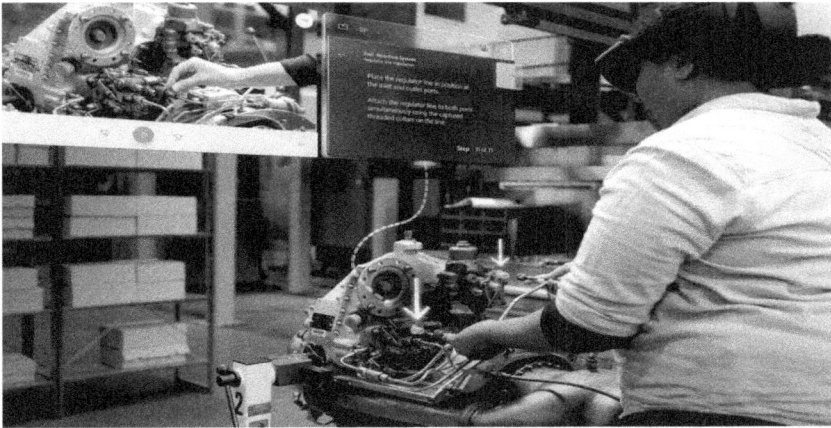

Figure 13.4: Microsoft Hololens

Source: Microsoft

My point is that although I4 will eliminate many jobs existing today, at the same time, it will also create unprecedented job opportunities for people who look into the future and acquire the right skills. Radical technologies aren't a new phenomenon. From steam engine to electricity, technology advances have been happening in the last 250 years. The birth of email might have caused fewer jobs for postmen, but, at the same time, it has created many jobs for IT administrators, technical support and software engineers. The Internet might have made a lot of newspapers and brick and mortar disappear, but, at the same time, it has created a lot of jobs in web design, data analyst and data engineers.

I4 and connected machineries will not be able to do the work on their own without interacting with humans. The low skilled job that will disappear will be replaced by new skilled jobs.

If you have read my book to this point, you will know there are plenty of opportunities out there if you can see the future. The world is full of problems, and each of them has a price tag attached to it. With a bit of ingenuity and action, perhaps one day, you will own your own version of I4 factory to solve these problems.

How smart is your city?

A smart city is a new urbanization model derived from I4 by applying new technology like ICT (Information and communication technologies), IoT, cloud computing, cyber securities, AI, big data for planning, and construction to manage the development of a city in a sustainable way. Every year, governments around the globe are racing to infuse new technology into every aspect of their operation. This includes public transportation, IT connectivity, e-governance, water, waste management, power supply, etc. Smart cities give government access to how the city is functioning in real-time.

As discussed in a previous chapter, IoT allows many network devices to communicate with each other over the Internet. With these IoT devices collecting data from people, vehicles, building to improve citizens' lives, from transport to healthcare to urban security, a smart city can identify the problems and opportunities in real-time and make better decisions based on data analytics and allocate resources more accurately.

One of these IoT technologies is smart street lamps. Smart street lamps adapt and dim when there is no activity, but brighten up when the sensor detects motion. According to the Chicago Department of Transport, smart street lamps will be 50% to 75% more efficient than traditional lights.[3] This will not only save energy costs but will add sensors like noise and seismic sensors, and have added functionality to detect threats to communities like gunshots and earthquakes. Besides lighting, smart cities also revolutionize the way a city manages waste. Smart waste handling, like an automated underground waste collection system (AWCS), is a network of underground-branched pipes that vacuum trash below ground with disposal inlets (e.g. smart bins) and terminals.[4] An RFID card reader is used to control and restrict system access and collect disposal data from users.

Virtual Singapore is perhaps the most distinctive example of how a smart city works. It is a dynamic 3D city model and collaborative platform that includes the 3D maps of Singapore.[5] You can visualize it as a SimCity game brought to life but with real-world properties. Scientists and urban planners can run simulations and experiments based on the rich data collected. It can predict the impact of natural disasters, and estimate the impact of traffic, pollution and other factors before urban development.

Figure 13.5: Virtual Singapore

Source: Singapore Government

I expect cities around the world will follow this type of model in the near future. With so much data underpinning smart cities, the sky is only the limit of what a Virtual City like this can potentially do.

Chapter 14
CERN

T he World Wide Web, as we all know, is one of the best-known innovations invented by Tim Berners-Lee from CERN in 1989.[1] It completely changed our life in the information age. But WWW is just one of the many CERN technologies in its portfolio; the organization's expertise in science, technologies and industry is vital in defining the future in the 21st century.

CERN is the European Organization for Nuclear Research, derived from the name *Conseil européen pour la recherche nucléaire*. Twelve countries in Western Europe established it on 29th September 1954.[2] Its goal is to unite talent from all over the world to push the frontiers of science and technology for the benefit of all.

Right now, it is conducting the largest, most expensive scientific experiment in the world.

The Largest Machine in the world

The Large Hadron Collider (LHC) is the largest and highest-energy particle collider in the world. CERN built it between 1998 and 2008, consisting of a 27 km ring of superconductor magnet with a number of accelerating structures to boost the energy of the particles along the way.[3]

Figure 14.1: Large Hadron Collider

Source: CERN

Inside the accelerator, two high-energy particle beams (e.g., protons) travel in proximity to the speed of light in opposite directions to collide with each other inside an ultrahigh vacuum tube. These beams are guided by a strong magnetic field, maintained by superconducting magnets. Electromagnetics are built from coils of special electric cables that operate in a superconducting state, efficiently conducting electricity with no resistance or loss of energy. To achieve this, it requires chilling the magnet to -271.3°C – a temperature colder even than outer space. That is why much of the accelerator is connected to a distribution system of liquid helium.

But what is the reason for building the world's largest machine to collide particles?

One of CERN's main goals of colliding particles is to answer fundamental questions, such as what all matter is made of. Understanding this will be a huge leap in mankind's knowledge in understanding the composition of our universe. It will be as remarkable as the 1969 Moon landing.

If you remember all the technologies we talked about in previous chapters, all of them are related to atoms made up of subatomic particles. The electricity we use every day is derived from electrons, which is a type of subatomic particle. Light is derived from protons, another type of subatomic particle. Our DNA is a molecule of atoms, which are made up of particles. Every matter around us, whether living or synthetic, is made from particles. So, understanding what elementary particles are made of and their properties will help us to understand the smallest building block of matter. And if you revisit Figure 2.1, an atom is made up of extremely tiny particles called protons, neutrons and electrons. Protons and neutrons are in the center of the atom. Electrons surround the nucleus. Protons have a positive charge. Electrons have a negative charge. Neutrons have no charge. This is as far as our high school chemistry goes regarding subatomic particles inside an atom.

But what is inside a subatomic particle like a proton?

At 10.28 AM on 10th September 2008, a beam of protons is successfully steered around the 27-kilometre Large Hadron Collider (LHC) for the first time.[4] To do that, scientists strip electrons from hydrogen atoms to produce protons. The protons enter LINAC2, a machine that fires beams of protons into an accelerator called a PS Booster that use devices called radiofrequency cavities to accelerates protons. These cavities have an electric field that pushes the protons to accelerate. Giant magnets produce the magnetic fields required to keep the protons on track. Once the right energy level is reached, the PS booster injects the protons into another accelerator called the Super Proton Synchotron (SPS). The SPS then injects these beams into the LHC, with one beam travelling clockwise and another travelling anticlockwise. LHC accelerates the beam of protons around the 27 km until they reach 99.9999991% of the speed of light. And when two protons collide successfully, they break apart into even smaller particles called quarks, which only exist for a brief moment.

The Standard Model

The Standard Model is perhaps the best model describing the subatomic world so far. The theory incorporates three of the four fundamental forces, excluding gravity.

In the standard model, all matter around us is made up of elementary particles in two basic types called quarks and leptons.

Physicists found that protons and neutrons are actually made up of quarks. A proton contains two "up" quarks (i.e., u) and one "down" quark (i.e., d). Neutron, on the other hand, contains two down quarks and one up quark.

If you look at the columns in the Standard Model, electrons, up quark and down quark seem to be all we need to build an atom. However, high energy physics discovered that there are actually six quarks – up, down, charm, strange, top and bottom. These quarks come with a variety of masses. The same was found for electrons, which have heavier siblings like muon and tau. These particles only exist for a very short moment in high-energy collisions and then decay into light particles. Such decay involves the exchange of force particles called W boson and Z boson, which have masses.[5]

Drei Generationen
der Materie (Fermionen)

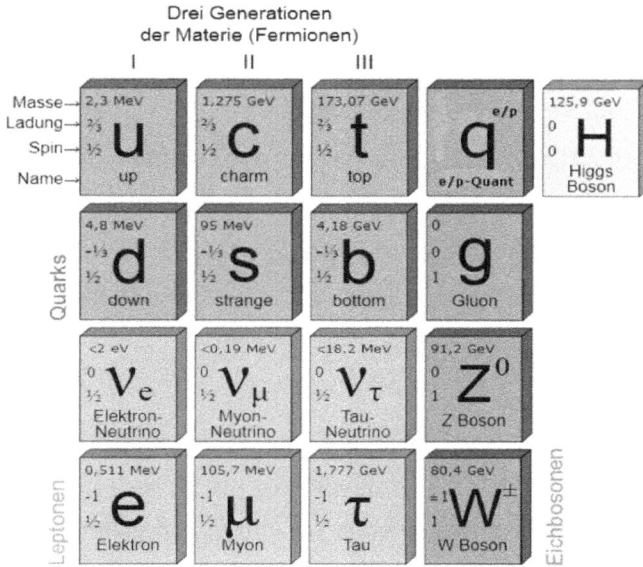

Figure 14.2: The Standard Model

Source: phys.org

[Note: There are four fundamental forces at work in the universe: the strong force, the weak force, the electromagnetic force, and the gravitational force. They work over different ranges and have different strengths. Gravity is the weakest, but it has an infinite range. If you drop a magnet, it falls to the ground because of gravity. However, you can lift that magnet by using another magnet because the force between two magnets overcomes gravity. The electromagnetic force also has an infinite range, but it is many times stronger than gravity. The weak and strong forces are effective only over a very short range and dominate only at the level of subatomic particles. Despite its name, the weak force is much stronger than gravity, but it is indeed the weakest of the other three. The strong force, as the name suggests, is the strongest of all four fundamental interactions.]

Higgs Boson - The God Particle

The final and perhaps most exciting particle in the Standard Model is the Higgs Boson, also known as the God Particle.[6]

On 4th July 2012, the ATLAS and CMS experiments at CERN's Large Hadron Collider (LHC) announced they had each observed a new particle in the mass region around 126 GeV. This particle is consistent with the Higgs Boson, but it will take further work to determine whether or not it is the Higgs boson predicted by the Standard Model.

On 8th October 2013, the Nobel prize in physics was awarded jointly to François Englert and Peter Higgs for the theoretical discovery of a mechanism that contributes to our understanding of the origin of mass of subatomic particles, which recently was confirmed through the discovery of the predicted fundamental particle by the ATLAS and CMS experiments at CERN's Large Hadron Collider.

So, why is Higgs Boson the God Particle?

You see, the Standard model isn't a complete model because it entirely misses out on gravity. Higgs Boson is so important to the Standard Model because it signals the existence of Higgs field – an invisible energy present throughout the universe that imbues mass in subatomic particles like quarks and leptons that make up ordinary matter.[7]

But how does the field give mass to particles?

Suppose you have a cocktail party filled with particle physicists at the CERN laboratory? The crowd of physicists represents the Higgs field. Imagine a biologist enters the party, and no one will talk to him. He could get across the room easily to get a drink without interacting with the crowd. This is the same way that some particles like photons do not interact with the Higgs field. They are massless. Now, suppose Albert Einstein entered the room. All the physicists will crowd around him.

Unlike the biologist, he interacts a lot with the crowd and moves slowly across the room to get a drink. Einstein is like one massive particle through the interaction with the Higgs field.

But, if the crowd of physicists is the Higgs field, how does Higgs Boson fit into this analogy?

Let's pretend the crowd is evenly spread across the room. Suddenly, someone pops their head in and spreads a rumor. People near the door will hear the rumor first. But people far away won't. They will move together to the people at the door to ask and create a clump in the crowd. As they have heard the rumor, they will return back to their original position to discuss it, and another clump will form, and so on. This clump is analogous to the Higgs Boson. It is not Albert Einstein interacting with more people in the room. It is the interaction with the crowd that causes it to gain mass. Mass comes from interaction with a field. A particle gets more or less mass, depending on how it interacts with a field. It is just like different people moving across a room at different speeds, depending on their popularity. Higgs Boson is like a clump in the field, like a rumor crossing the room.

***

CERN had advanced the frontiers of technology and brought nations together through science. Its contribution to the kind of knowledge not only enriched humanity but provided the wellspring of ideas that will become the technologies of the future.

Just last year, CERN released a design report for the Future Circular Collider, which would be four times as big as the current LHC.[8]

I often ask, why do we continue to run experiments, smashing together protons, if we've already discovered the Higgs Boson?

The truth is, there is still a lot we do not know about our universe. There are a number of questions that the standard model does not answer. For example, studies of galaxies and other large-scale structures in the universe indicate that there is a lot more matter out there than we observe. We call this dark matter because it is not visible to the naked eye. The most common explanation to date is that dark matter is made of an unknown particle. Physicists hope that the LHC of the future may be able to produce this mystery particle and study it. That would be an amazing discovery.

Chapter 15
The Quantum Era

It is almost impossible to picture how weird things can get down to the smallest of scale.

In the quantum realm, particles don't like to be tied down to just one location or follow one path. It is almost as if things can be in more than one place at the same time. Imagine if people can behave like the particles inside the atom, most of the time, you don't know exactly where you are. They could be anywhere until you look for them.

As bizarre as it might sound, scientists have been using them to predict how atoms and particles behave. Strangely enough, after countless experiments, quantum laws have always been right.

We called it quantum mechanics.

How did we discover quantum mechanics?

God does not play dice with the universe

-Albert Einstein

Not long ago, we thought we had all the laws of physics pretty much figured out. Whether it is the fall of an apple or ripples across the surface of water, the law of physics allows us to predict the behavior of things with certainty. It all makes common sense.

It was one hundred years ago when scientists struggled to explain the unusual properties of light – the kind of light that glowed from gas when they were heated in a glass tube. Scientists observed this light through a prism and saw something interesting – it forms distinct lines at very specific color.

In 1913, Niels Bohr, a Nobel Prize winner in physics, gave an analogy of how atoms resembled tiny solar systems with subatomic particles like electrons orbiting around the nucleus, much like planets orbit around the sun.[1] But, unlike the solar systems, electrons do not travel in a fixed orbit. When heated, electrons could leap from one orbit to another. Each leap would emit energy in the form of light in a very specific wavelength. That is why atoms produce a very specific color. This is where we got the phrase the *quantum leap*.

What makes a quantum leap so special is that electrons can leap from one random position of an orbit to another random position in another orbit without traveling in space. Imagine you can be in Australia this moment, and the next moment, you are in New York City. It is a very different set of rules we see in our everyday life.

The famous double-slit experiment further proves that light displays the fundamental probabilistic nature as both particles and waves.

In fact, the fundamental nature of reality at the deepest level is determined by a probability wave. Erwin Schrödinger, a Nobel Prize winner in physics, developed an equation to describe this wave function or state function of a quantum-mechanical system called the *Schrödinger equation*.[2] In the double-slit experiment, the result of the *Schrödinger equation* predicts the likelihood of a light particle existing in a particular space with great certainty.

$$\frac{-\hbar^2}{2m}\nabla^2\Psi(r) + V(r)\Psi(r) = E\Psi(r)$$

However, Albert Einstein found this concept rather counterintuitive. He believes that God does not lay dice with the universe. Even so, it is undeniable that the equation of quantum mechanics gave scientists the power to predict the behavior of groups of particles and tiny particles with high precision. It is this power that sparked the quantum revolution that led to many very big inventions today, like a laser, transistors, and IC. The equations of quantum mechanics help engineers to design microscopic switches that direct the flow of electrons that control our devices every day.

Probability and the Act of Observation

I like to think the moon is there even when I'm not looking.

-Albert Einstein

But, in spite of the success of quantum mechanics, many scientists remain deeply mysterious about it. Albert Einstein once raised a question involving probability and measuring the observation.

Niels Bohr, a Danish physicist, believes that before you measure or observe a particle, its characteristic is uncertain. It is not until the moment you observe it, and only at that point will the location of uncertainty disappear. Niels accepted that the nature of reality was inherently fuzzy.[3] Einstein, on the other hand, believes that certainty is not just measured or looked at, but it is there all the time, just like the moon is always there even when we are not looking at it. Einstein was convinced that something was missing from the quantum theory. In his view, the detail of the location of particles should not depend on the observer.[4]

Quantum Entanglement

The ridiculous prediction that quantum mechanics makes is something called entanglement. Entanglement is a theoretical prediction that comes from the equation of quantum mechanics. Here is an analogy.

Imagine two particles become entangled if they are close together, and their properties become linked. Quantum mechanics said if you separate those particles, their properties remain entangled, even over long distances.[5]

To understand how profoundly weird this is, consider a property of electron called spin, as shown in the last chapter. Like other quantum qualities, an electron spin is fuzzy and uncertain until the moment you measure it. It can spin close-wise or anticlockwise.

Now, imagine the spin of an electron as a spinning wheel that can either land on blue or red, and you have two of these wheels. When one lands on blue, the other is guaranteed to land on red, no matter how far apart they are. In terms of particles, when you make a measurement of the state of a particle, you would affect the state of its entangled partner no matter the distance.

There is nothing connecting to them. No wires. No cables.

Isn't that spooky?

This is quantum entanglement.

Einstein doesn't believe in this. Instead, he gave an analogy that the particles are like a pair of gloves. Imagine if someone separated the two gloves and put each in a case. One of the cases goes to the Moon, and the other one stays on Earth. When people on Earth open the case and find a left glove, he will automatically know that the other one on the Moon is the right glove. Einstein applies the same phenomenon to quantum entanglement. Whatever configuration the electrons are in must have been predetermined the moment they are separated.

So, which theory is right?

Is it Niels Bohr's analogy that quantum particles are really communicating through spooky action like the matching spinning wheels? Or is it actually nothing spooky like Einstein said, and the state of the particles were already predetermined like a pair of gloves?

In 1964, a physicist called John Bell solved this problem. The Bell test has its origins in the debate between Einstein and other pioneers of quantum physics, principally Niels Bohr. In his paper, Bell concluded that the quantum wave function does not provide a complete description of reality. There are some hidden random variables that are behind this randomness of the particles. Bell turned the debate into an experimental question by measuring entangled particles and comparing their spin in many different directions.[6]

After many repeat experiments by different scientists, the shocking result is that they have proven that the math to quantum mechanics is actually correct. Entanglement is real. Quantum particles can be linked across space. Measuring one particle can, in fact, instantly affect the state of its entangled partner.

Up to this point, you may be wondering what quantum entanglement has to do with the context of this book? How does quantum entanglement hope to have any real-life application?

Please read on.

Quantum Teleportation

Teleportation is a common subject in science fiction films like Star Trek. The idea is to transport things from one place to another instantly without crossing space in between.

But will teleportation ever become a reality? How is it related to quantum mechanics?

In the Canary Islands off the coast of Africa, scientists performed quantum experiments to use quantum entanglement to teleport tiny particles like photons in a lab on an island of La Palma. One entangled photon stays on La Palma, while the other one is sent by laser to the island of Tenerife about 89 miles away. A third photon, the one the scientists want to teleport, interacts with the entangled photons on La Palma. The team studies the interactions by comparing the quantum states between two particles. The amazing part is that because of this spooky action, the team is able to use that comparison to transport the entangled proton on the distant island into an identical copy of the third photon. The information contained in the third photon's quantum state is transmitted from the photon on La Palma to the photo on the island of Tenerife through quantum entanglement - without actually travelling the intervening distance. It's as if the third photon had teleported across the sea without travelling the space between the islands![7]

But could this technology go further?

Since we are all made up of particles, could this process make human teleportation possible one day?

Imagine we have a teleportation machine, which has a chamber of particles in Sydney, that's entangled with another chamber in New York. Let's say anyone in Sydney wants to try out teleportation can step inside a pod. An operator in Sydney will then use the device to scan the huge number of particles in our body in the pod. It is jointly scanning the particles in the other chamber and creates a list that compares the quantum state of the two sets of particles. Through the spooky action of quantum entanglement, a copy of the state of particles that represent me will be sent to the operator in New York. He will then use the data to reconstruct the exact quantum state of each and every one of the particles in my body in the chamber of New York and create a new me.

This might be how teleportation will work in the future.

Remember, it is that entanglement that allows my quantum state to be extracted from Sydney and then reconstructed in New York. However, during the process, it will destroy the original me.

Whether or not teleportation will arrive in the future, the fuzzy uncertainty of quantum mechanics has a lot of potential applications.

One of them is the quantum computer.

The Birth of the Quantum Computer

A major milestone took place in the history of mankind since we entered the 21st century. All the revolutionary digital technologies that happened are linked to one key trend called Moore's law. Shrinking down transistors had powered 50 years of advance in computing. As long as Moore's law holds, breakthroughs in cellphone and computing technologies continue.

In the past, Intel took roughly 18 months to double the number of transistors on an IC. However, in 2020, Moore's law is reaching technological barriers to continue shrinking down the number of transistors and fit in an IC. Sooner or later, Moore's law will reach its limit.

Will the end of Moore's law halt computing's exponential rise?

While the death of Moore's law has been often predicted, the era of quantum computing is just on the horizon.

A quantum computer isn't just a more powerful version of the computer we use today. It is something else entirely.

To illustrate this, imagine you are in a Las Vegas casino, and you decide to play a game on one of its computers. The computer can make a move in the game, just as a human player would. Say, for example, it is a coin game. It starts with a coin showing a head, and the computer will play first. It can choose to flip the coin or not, but you don't get to see the outcome. Next, it is your turn to play. Like the computer, you can choose to flip the coin or not. Your move will not be revealed to the computer, and vice versa. After three rounds, the coin's state is revealed. If it is a head, the computer wins, and if it is tail, you win.

In a simple game like this, you have a 50/50 chance of winning.

This is not the case if you play against a quantum computer.

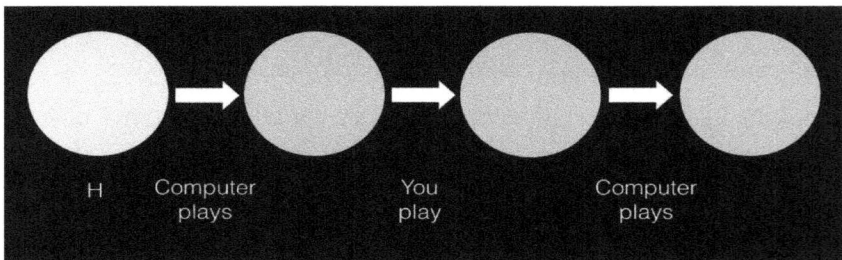

Figure 15.1 : Flipping coin

Source: Author

A quantum computer operates by controlling the behavior of fundamental particles like electron and photons.. This is completely different to how a regular computer works. It is like an electric light bulb is not a more powerful version of candle. If the same game was played against a quantum computer, it might surprise you that the quantum computer has a 97% chance of winning

So, how can it achieve such amazing results?

Here is how it works.

A regular computer simulates heads or tails of a coin as a bit – head is 1, tail is 0.

A quantum computer, on the other hand, uses a quantum bit that has a more fluid, no binary identity. It exists in a superposition, with some probability being 0, and some probability being 1. It is like a spectrum. It can be 60% 0 and 40% 1, or 20% 0 and 80% 1. The possibilities are endless. To understand this, we have to give up our precise definition of 0 and 1 and allow some uncertainty.

Let's revisit the game again.

The quantum computer creates a fluid combination of heads and tails, zero and one. So, no matter what the player does, flip or no flip, the superposition remains intact. Think of it as stirring a mixture of two fluids.

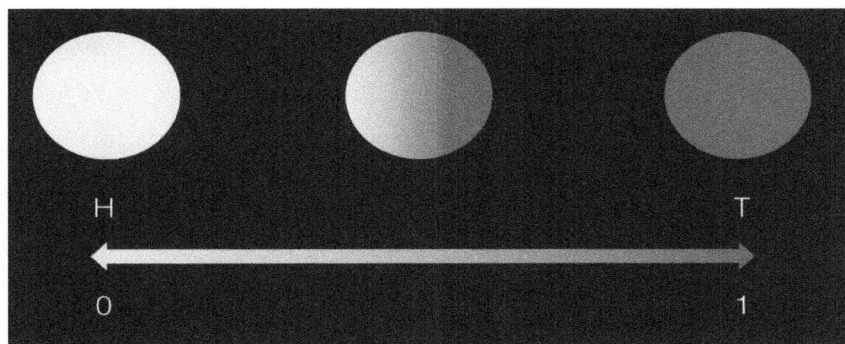

Figure 15.2 : Quantum Bit

Source: Author

Whether you stir or not, the fluids remain in a mixture. But in the final move, the quantum computer unmixes the 0 and 1, perfectly recovering heads so that you lose every time.

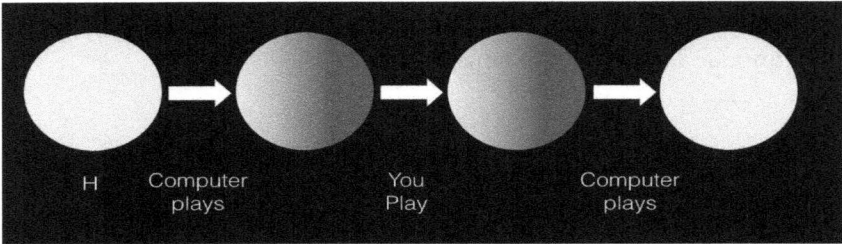

Figure 15.3 : Quantum Computer wins

Source: Author

If you think it is a bit weird, it is understandable. Regular coins do not exist in a combination of heads and tails. We do not experience this fluid reality in our everyday lives. But even though we do not experience quantum strangeness, the quantum computer wins because it harassed superposition and uncertainty. These quantum properties are powerful for building future quantum technologies.

Quantum computers could be used to create private keys for encrypting messages sent from one location to another. Hackers cannot secretly copy the keys perfectly because of quantum uncertainty. To hack the message, they would have to crack the law of quantum physics to hack the key. This kind of unbreakable encryption technology will be very useful for banks, military and other institutions.

Apart from that, quantum computers can be used to transform healthcare and medicine. It can be used to design and analyse molecules for drug development, which is impossible for the conventional computer.

Besides that, quantum computing can be used to teleport information from one location to another without physically transmitting the information. Sounds like science fiction? But it is possible. The fluid identities of quantum particles get tangled across space and time. If you change something about one particle, it can impact another. This creates a channel for teleportation. This could be the beginning of a framework of a future quantum Internet and a new digital era.

***

A conventional computer uses binary code in terms of bits, zero or one. A bit is the smallest unit to represent information. A quantum computer communicates in quantum bits or qubits.[8] This is a bit of information that can be a fuzzy mixture of 1 and 0, just like an electron can be a fuzzy mixture of spinning clockwise or anticlockwise. So a qubit can multitask. That is why a quantum computer can do computation in ways a conventional computer can only dream of doing.

In theory, a qubit could be made from anything that acts in a quantum way like an electron or atom. The qubits at the heart of a quantum computer are just tiny superconductor circuits built with nanotechnology that can run in two directions at once. Since qubits are so good at multitasking, we can increase our computing power exponentially if we can figure out how to get qubits to work together.

A simple analogy would be to imagine you are trapped in a maze, and you want to find a way out. There are so many possibilities, but you can only try one route at a time. You will meet a lot of dead ends and make many wrong turns. This is how a traditional computer solves a problem today. A quantum computer, on the other hand, works differently. Instead of trying one route at a time, it tries all possible routes at the same time. Since a particle can be in many states at once, a quantum computer could investigate a huge number of paths or solutions stimultaneously. That is why a future quantum computer will allow us to forecast weather, predict natural disasters and solve problems that would be almost impossible today without enormous amounts of computation power. A quantum computer can get the job done with just a few hundred atoms, so its brain might be smaller than a grain of sand.

Where Quantum Mechanics Leads?

So, where will quantum technology lead us in the 21st century?

So far, no one has explained why the quantum fuzziness disappears as things increase in size. As powerful and accurate as quantum mechanics has proven to be, scientists are still trying to find out the reason. Are there some details missing in the equations of quantum mechanics? Is it because despite so many different possibilities in our world, the missing details would adjust the uncertainty as we increase in size from atoms to objects? Hence, the result is a single certainty. Some scientists believe that each and every possibility actually never diminishes. They believe the possibilities are always there and exist as a parallel universe to our own. And our reality is much grander than we thought it is. This is the frontier of quantum mechanics. So far, no one knows where this will actually lead.

Chapter 16
A Time Traveller's Gift

If you pick up an apple and mark A at the front and B at the back, what is the fastest route you can travel from A to B in that three dimensional space?

Figure 16.1: Apple

Source: Author

Time travel is perhaps one of the most interesting topics in science. The idea of moving between different points in time has been the topic of science fiction for decades.

Truth be told, we are all time travellers. We are all traveling forward in time at the same rate, one hour forward 24 hours a day. But, the question about time travel is not about whether we can travel in time. It is about changing the rate of travel forward and backward in time.

Special Relativity

The distinction between the past, present and future is only a stubbornly persistent illusion.

-Albert Einstein

For decades, Albert Einstein had been troubled by the fact that Newtonian mechanics and Maxwell's equations, the two pillars in physics, are incompatible. One night, when Einstein boarded a tram home, while retreating from the Zytglogge clock, he imagined himself riding home in a streetcar in Bern. He saw the clock tower passing behind him and wondered how the clock would appear as the streetcar moved faster and faster.

At 300,000 km per second, the streetcar would move at the speed of light. The tower's clock would appear frozen to Einstein in the streetcar, but the watch Einstein wore would tick normally. Einstein concluded that the faster you moved through space, the slower you moved through time. This concept is known as special relativity. Time and length are not as absolute as our everyday experience would suggest. Moving clocks run slower, and moving objects are shorter, as they are moving close to the speed of light.

So, like Stephen Hawking said, if you fly off at an incredible speed in a spaceship and returned back to Earth, less time would pass for you than it would for anyone you left behind.

Gravity Talk

If you have watched the movie Interstellar, you will probably wonder how one hour on planet Miller equals seven years pass on Earth. This explanation comes down to Gravitational Time Dilation – an effect that measures the amount of time elapsed between two events as measured by observers at varying distance from a gravitational mass.

In Newton's theory, space and time are absolute. They are fixed independent quantities. Time passes uniformly, regardless of what happens in the world. But for Einstein's picture, three-dimensional space and time are linked as a single quantity – spacetime. Imagine spacetime as a fabric sheet that can be curved by the presence of gravity of mass.

Time goes faster, the farther away you are on the earth's surface because the stronger the gravity, the more spacetime curve blends, and the slower time runs. When you board an aircraft, you will see that an atomic clock ticks slightly faster than a clock on the ground. The reason is that gravitational force decreases with altitude; it is inversely proportional to the square of the distance from the center of the Earth. In fact, the classic example of how satellites have revolutionized research in relativity is the Global Positioning System (GPS). GPS has become an essential navigation tool in our daily life. Everything from public transport navigation to cruise missiles depends on it. It requires precision of about 14 nanoseconds. Special relativity predicts that on-board atomic clocks on the GPS fall behind clocks on the ground by 7 microseconds per day because of the slower ticking rate due to time dilation of their relative motion.[1] Imagine what would happen if our GPS did not have these corrections due to special relativity in place? The whole GPS system will become very unreliable.

So, the fall of an apple is our familiar worldview of the world. It is how we experience reality. But is it really because of the gravitational pull the Earth exerts on the apple, or are gravitational effects actually the result of a distortion in the fabric of spacetime? It looks like the gravity that we experience every day is not really how the world is actually structured.

Can we Time Travel?

Time travel is possible if you are focusing on travelling into the future. If you want to leapfrog into the future and see what Earth will be like a century from now, it is entirely possible.

Imagine you have a spaceship that travels to space at a proximity to the speed of light for six months, and you turn around and come back to Earth for six months? You have time travelled. Your time in space is elapsing much slower than the time on Earth. So, when you step off the spaceship on Earth, you are one year older, but the Earth has been through many, many years, depending on how close you travelled at the speed of light. This is one of the ways you increase your rate of travel forward in time.

There is also another method, which is harassing the black hole as a time travel engine.

Using Black Hole as a Time Travel Engine

As we all know, a black hole is a region in space where the gravitation pull is so great that even light cannot escape. That is why the object appeared black to the outside world, like a hole. In a black hole, gravity crushes all mass to one tiny point, which can be the size of an atom only. This point is called a singularity. The boundary of a black hole is called an event horizon. If something is inside the event horizon of a black hole, it will never get out.

But how is the black hole related to time travel?

Imagine you have a spaceship, and you travel at a proximity to the event horizon of a black hole? Without being sucked in, time will elapse more slowly for you than back on Earth. When you travel back to Earth, like before, it will be many, many years into the future, depending on how close you are to the event horizon and how long you have been travelling.

These are two ways to travel into the future.

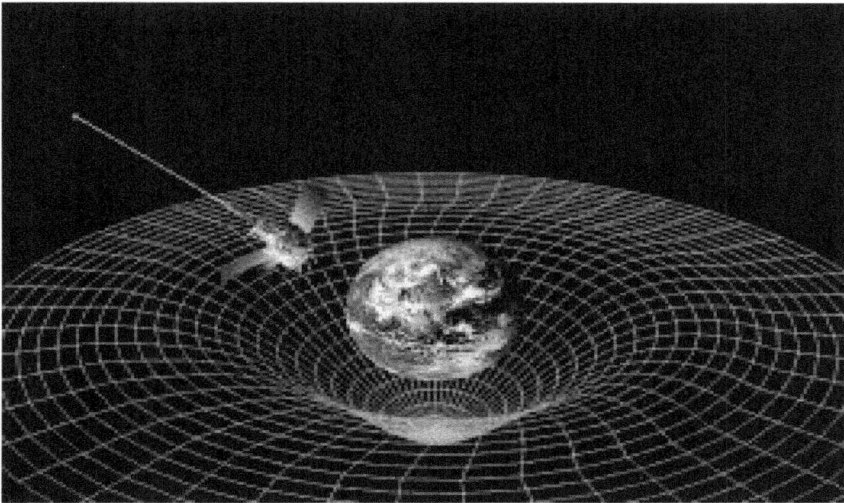

Figure 16.2: Time Travel at the Event Horizon

Source: NASA

To illustrate how it works, let me give you an example.

Imagine you start your time travel from Earth on 1st January 2020, and you plan to travel to a distant blackhole, and a round trip, and then come back to Earth.

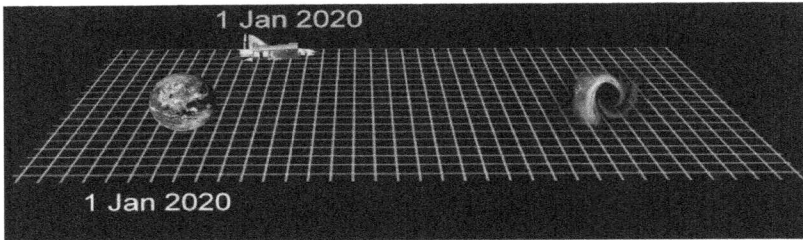

Figure 16.3: Time Journey begins on Earth

Source: Author

Because time runs slower for you than the ones you left behind, by the time you reached the proximity of the event horizon of the black hole, the time on your atomic clock on the space shuttle would be say, 31st July 2034, and the Earth time would be 20th November 2036.

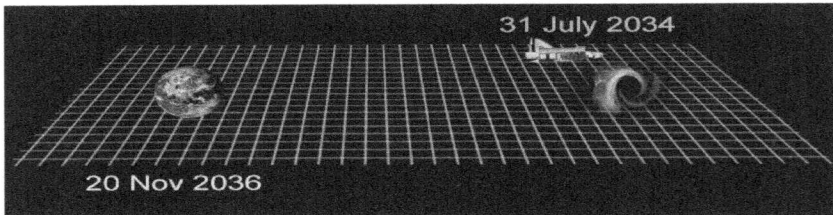

Figure 16.4: Time before Space shuttle made a round trip

Source: Author

Now, since time runs slower where gravitational force is the strongest, by the time you make a round trip near the event horizon, the atomic clock on your space shuttle would run significantly slower than Earth time. Like Einstein said, time elapses more slowly in the presence of strong gravity. So, after you made a round trip, the time on your atomic clock might be 26th Nov 2035, whereas Earth time is 28th Dec 2060.

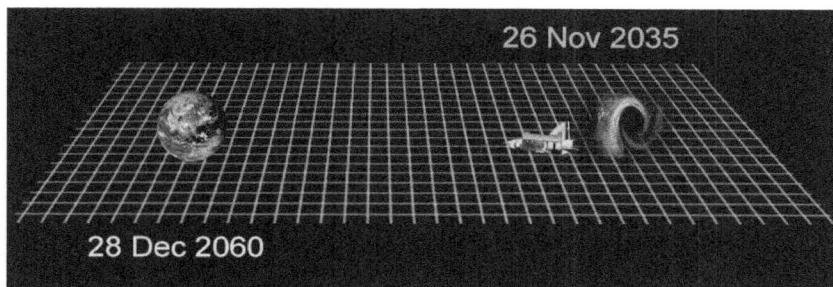

Figure 16.5: Time after Space shuttle made a round trip

Source: Author

By the time you return to Earth, the world will be a different place. You will find your friends thirty years older than you.

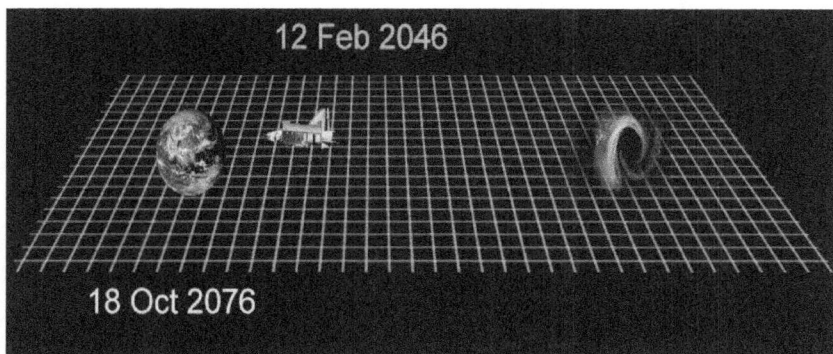

Figure 16.6: Home sweet home

Source: Author

This is basically what happens when you time travel to the future. But what about travel back to the past?

Solving the Grandfather Paradox

Perhaps might have heard of the Grandfather paradox of time travel – if time travel is possible, and you went back and prevented your grandparents from meeting, you would prevent your own birth and your subsequent time travel. Any actions that alter the past will have a contradiction to classical reasoning.

Logically speaking, travelling backwards in time is impossible. A time paradox is enough to draw a conclusion of the possibility of traveling back in time. After all, it appears to be a paradox instinctively unless disproved by science.

In 2014, Dr. Martin Ringbauer and his team of researchers at the University of Queensland simulated what would happen if a single photon were to be caught in a Deutsch "closed time like curve" (CTC) – causal loop in space-time that returns to the same point in space and time. They discovered that in the quantum regime, these paradoxes could be solved.[2] The fuzziness of quantum states prevents a paradox from happening. The photon is already in a quantum superposition of combined being existence and non-existence. In the classical state, you can either exist or not, but in the quantum realm, both states can happen.

Wormhole is a Hypothetical Time Travel Engine

The main dispute about the subject of time travel is whether we can travel back in time. Conventionally, physicists dismissed the idea of time travel because of the mentioned time paradox.

In 1935, Einstein and physicist Nathan Rosen used the theory of general relativity to propose an idea of the existence of "bridges" through spacetime.[3] For a simplified explanation, visualize a folded 2D spacetime fabric, where light travels from one point, say the entrance at the bottom of the plane, to another point, say, the exit at the top of the plane, along the curvature. Einstein imagined a wormhole as a shortcut between the two points. In other words, you can either travel from the entrance to the exit by following the long curvature or go the shortcut through a wormhole.

But how does travelling back in time in a wormhole work exactly?

Imagine that this time, we rerun the same experiment with our space shuttle travelling to the black hole, but this time, in the presence of a wormhole.

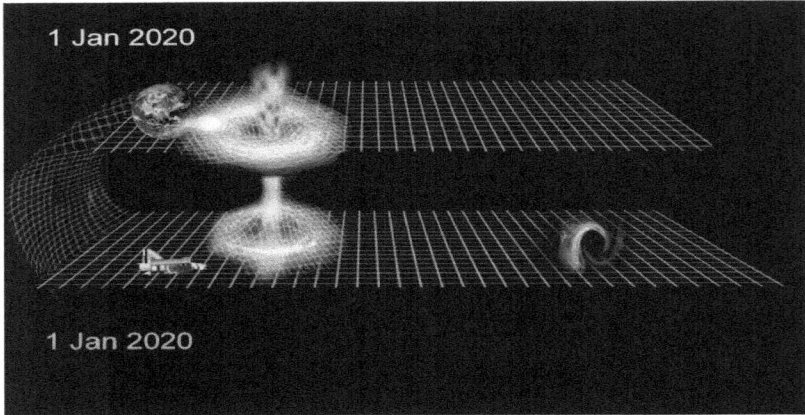

Figure 16.7: Time Journey begins on Earth again

Source: Author

The space shuttle starts on the 1st January 2020. Time elapsed more slowly as it took off.

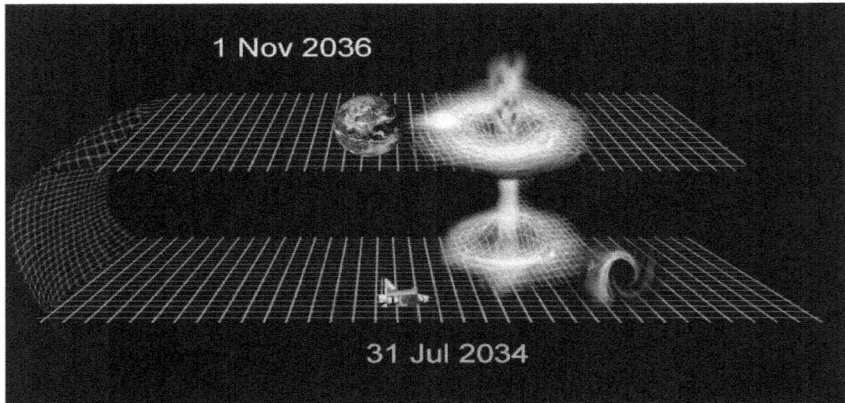

Figure 16.8: Time before Space shuttle made a round trip

Source: Author

Like before, when your space shuttle reaches the event horizon, the atomic clock on your space shuttle will begin to run significantly slower than Earth time.

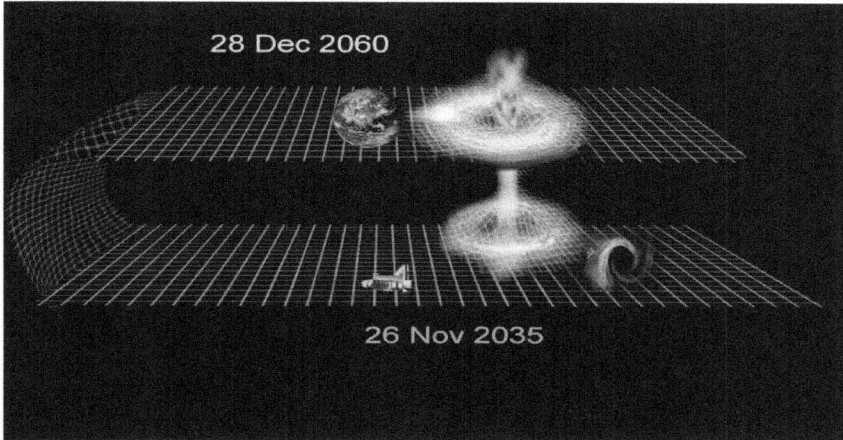

Figure 16.9: Time after Space shuttle made a round trip

Source: Author

By the time it gets back to its starting point, like before, there is a 30-year time difference. Notice that the wormhole hasn't played any role in here yet. It just went along for the ride.

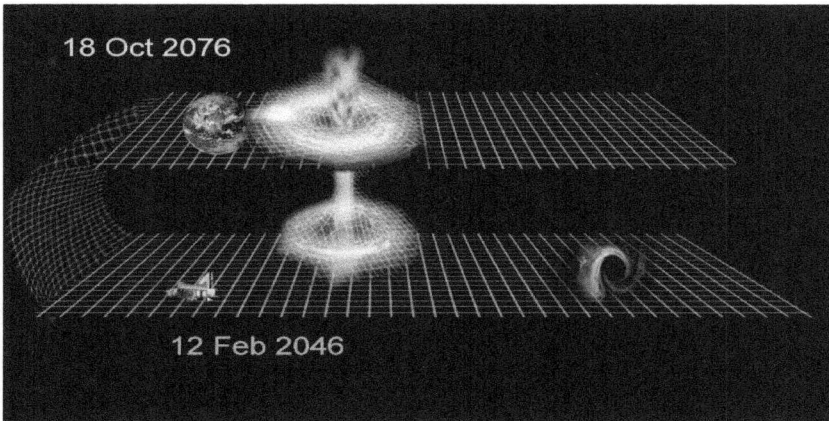

Figure 16.10: Home Sweet Home

Source: Author

The interesting part is here.

We rerun the exact same situation with our space shuttle travelling to the proximity of the black hole. This time, instead of viewing through the perspective of the space shuttle, we look at the situation from the perspective of the wormhole. The math shows that the two clocks at each opening of the wormhole agree. There are no time differences between them.

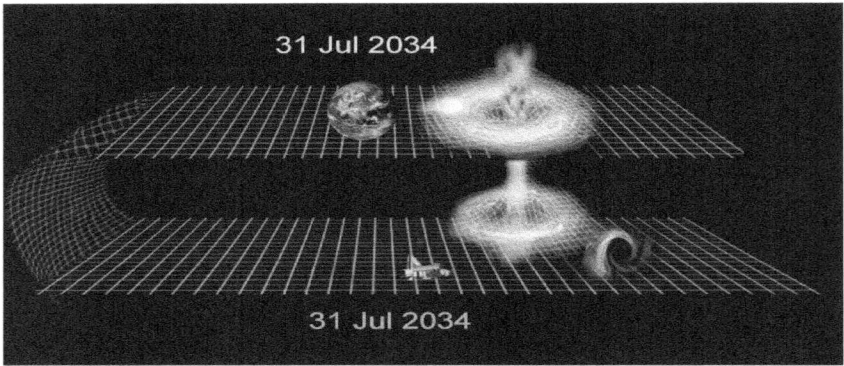

Figure 16.11: Clock view through both openings of the wormhole near the proximity of the blackhole

Source: Author

And when the space shuttle travels back to Earth, the clocks at each opening of the wormhole will continue to agree with one another.

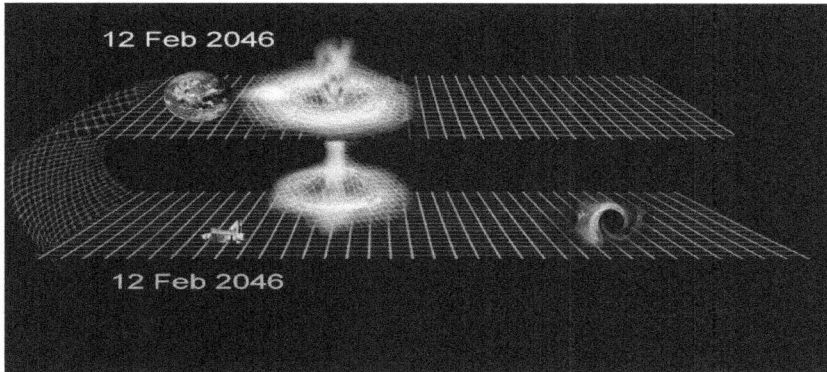

Figure 16.12: Clock view through both openings of the wormhole when space shuttle

returns to Earth

Source: Author

So, if you examine the same situation by comparing Fig 16.10 and Fig 16.12, you will discover that from the perspective of Fig 16.10, there will be a thirty-year time difference between the space shuttle and Earth. But from the perspective of Fig 16.12, there are no time differences at all between the two.

The truth is that both situations are correct, but just at different perspectives.

The first perspective is by going through the long route. The second perspective is by going through the shortcut. This is how a wormhole allows time travellers to travel to the past. Instead of traveling to 2076, a wormhole takes you 30 years back to the year 2046.

Assume this type of wormhole exists. This type of time travel cannot bring you back before 2020. It just brings you to a less distant future.

So far, there is no observational evidence that a wormhole exists. Even if it does, it will be very unstable. Anyone who tries to time travel through a wormhole will be crushed, as the wormhole will self-destruct as soon as it is formed. A fictional traversable wormhole can only exist if its entrance and exit were held by naturally occurring exotic matter with high enough density.

Chapter 17
Space Travel

Konstantin Tsiolkovsky (1857 – 1935), the Russian father of rocketry one said, *Earth is our cradle, but we cannot live in the cradle forever.*[1] In 1903, he published the famous rocket equation, which allows anyone to determine the relationship between weight, fuel and escape velocity – the speed required to escape the gravity of Earth.[2]

Despite the fact that Tsiolkovsky couldn't convert his vision into prototypes, he mapped out the theoretical basis of space travel that inspired future rocket scientists to make it a reality. One of them is Robert H. Goddard (1882 – 1945), the American father of modern rocket propulsion.[3]

On 19th October 1899, at the age of 17, Goddard had just read H.G. Wells' classic science fiction *War of the Worlds* and was inspired by spaceflight. He dreamt of a future that rockets could allow humanity to travel across the universe.[4] That young boy dedicated to make his dream come true. By 1926, Goddard had successfully constructed and tested the first liquid fuel rocket. Every year, every 19th October, Goddard would have a private holiday to celebrate his greatest thought: the possibility of breaking free from gravity to travel to other planets. The NASA Goddard Space Flight Center was named after him.[5]

The Great Space Race

V-2 missile (Vengeance Weapon 2) was the first practical Intercontinental ballistic missile (ICBM) developed by German scientist Wernher Von Braun in 1936.[6] Braun designed it based on Tsiolkovsky and Goddard's sketches. V-2 missile travels 3580 miles per hour, 4.6 times faster than the speed of sound. It dwarfed all previous achievements in the history of rockets in terms of speed and range.[7] It was the first rocket in the world that successfully left the boundary of the atmosphere and entered space. It made no warning noise until impact and was impossible to defend against it. It was a weapon used by Nazi Germany to terrorize England during WWII.[8]

When Germany was defeated after WWII, the Allies force – the U.S., the UK, and Soviet Union raced to capture the key manufacturing sites and technology of V-2 rockets. Braun surrendered to the U.S., and the U.S. smuggled enough hardware to build its own missiles. The Soviet Union took over Germany's V-2 manufacturing site, captured a few V-2 engineers and re-established the production.

During WWII, the U.S. and the Soviet Union fought together as Allies against the Axis power. However, after the war, their relationship became complicated. The U.S. feared the Soviet Union's domination of Eastern Europe and their intention to spread communism worldwide. This sparked the beginning of a Cold War between the two superpowers.

On 4th October 1957, the Soviet Union launched the world's first artificial satellite – *Sputnik 1*.[9] This was a huge blow to the U.S. prestige. It spread panic in the U.S. Knowing that the Soviet Union was leading the world in space technology started the Space Race. *Explorer 1* was the first satellite the U.S. launched on 31st January 1958.[10] At the same year, The National Aeronautics and Space Administration (NASA) was created.[11] President John F. Kennedy called for a program to put a man on the Moon by the end of the decade. First, NASA launched Project Mercury (1958 – 1963) – the first human spaceflight to put man into Earth orbit and return him safely.[12] Project Gemini (1961 - 1966) was launched with two astronauts to understand an astronaut's ability to fly a long duration mission and how spacecraft could dock in orbit around Earth and the Moon. It was an important lessons for space travel.[13] Project Apollo (1961 – 1972) was the third spaceflight that successfully landed man on the Moon.[14] On 20th July 1969, Apollo 11 had fulfilled President John F. Kennedy's ambition and was the first spaceflight that landed humans on the Moon. When Commander Neil Armstrong planted the first human footprint onto another world, he proclaimed: *That's one small step for a man, one giant leap for mankind.* Saturn V was the rocket built for the Apollo mission to send people to the Moon.[15]

Figure 17.1: Moon landing

Source: NASA

The Soviet Union, on the other hand, had ceded the Space Race because of a series of disasters. There is no single, easy explanation.

Besides running out of money, one of the reasons is because Korolev, who was in control of the Soviet Union's rocket, died in 1966.[16]

Without the Space Race with the Soviet Union, the U.S. quickly lost momentum on the space program. The astronomical cost of the space program became unsustainable. Beside the Space Race, President Lyndon Johnson declared war on poverty in 1964. In 1965, the U.S. entered the Vietnam War. Worst of all was the tragedy of the *Challenger* explosion in 1986[17] and the *Columbia Disaster* in 2003[18], which further weakened public support. All these events meant less funding for the space program and began its decline. In 2010, President Obama cancelled the Space Shuttle program and Constellation program.[19] That is why we haven't heard much about space exploration by NASA in recent decades.

But the contest of the Space race didn't just end here.

In 1975, the U.S. and the Soviet Union worked together on the Apollo-Soyuz Project – a mission involving both nations to dock their capsule in space together.[20] The mission began when two cosmonauts launched the Soyuz 19 capsule. A few hours later, the Apollo capsule followed and docked with it. This marked the beginning of an era of international collaboration. In 1998, the largest collaboration in space began – the International Space Station (ISS). The ISS consists of Canada, Japan, the Russian Federation, the U.S., Belgium, Denmark, France, Germany, Italy, The Netherlands, Norway, Spain, Sweden, Switzerland and the United Kingdom. This enormous achievement marks how much can be achieved when the world comes together.[21]

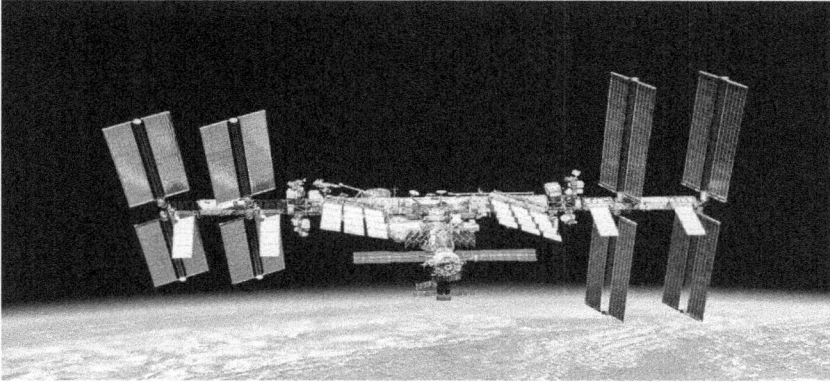

Figure 17.2: International Space Station

Source: NASA

However, not every nation is allowed to take part in the ISS. China has never take part in the program and have been officially banned since 2011. But China put forward their own space program. In 2003, it was the only nation beside the U.S. and Russia to launch a human to space.[22] Despite China continuing to pursue to be a member of the ISS, the U.S. Congress eliminated the possibility of China joining the ISS through legislation called the *"Wolf Amendment"*.[23] This prevents any U.S. companies and NASA from sharing technological advancement with the Chinese government. This sparked a new Space Race in the 21st century.

In 2003, China founded the China National Space Administration (CNSA) for national space program and development of space activities.[24] Despite being banned from the ISS, China rapidly advanced their space program. On 24th October 2007, China launched the Chinese Lunar Exploration Program called the *Chang'e* program after the Chinese moon goddess *Chang'e*.[25] *Chang'e-1* surveyed the lunar soil for the element Helium-3, which could one day power nuclear reactors. In 2011, they launched their own space station known as *Tiangong 1*, which was soon followed by *Tiangong 2* in 2016.[26] In 2019, China became the first nation to successfully land a spacecraft on the other side of the Moon.[27]

The Golden Age of Space Exploration

The whole idea is to preserve Earth.. The goal is to be able to evacuate humans. The planet would become a park.

-Jeff Bezos

Only recently, NASA has been able to reclaim the leadership in space exploration. In November 2015, NASA reaffirmed the goal of sending humans to Mars.[28]

Besides NASA, a lot Silicon Valley entrepreneurs with private and public funding also jumped into space exploration. Jeff Bezos, the founder of Amazon, was one of them. He was the richest man in the world in 2017. Bezos has been interested in space from a very early stage. His dreams of spaceflight were fostered at the age of 5 when he saw NASA astronaut Neil Armstrong take humanity's first steps on the Moon in 1969. In 2000, he founded his own company *Blue Origin* to build his own launch pad and rocket system called *New Shepard*.[29] The name *New Shepard* is named after Mercury astronaut Alan Shepard - the first American to go to space. It is a reusable rocket designed to take astronauts and research payloads past the Kármán line – the internationally recognized boundary of space.[30] At that height, any passengers on board would experience a few minutes of weightlessness before coming back to the planet's surface. Apart from that, Blue Origin's New Glenn is a remarkable innovation that will build the road to space. It is able to carry people and payloads routinely to Earth's orbit and beyond, even on a moving ship.[31] *Blue Origin* designed operationally reusable rockets, with demand high performance engines capable of deep throttling for soft landings.[32]

In 2017, Bezos announced a short-term plant for *Blue Origin* to set up a delivery system on the Moon. He would sell $1 billion worth of Amazon stock every year to help fund *Blue Origin*.[33] Apart from that, he presented the *Blue Moon* spacecraft. *Blue Moon* can deliver multiple metric tons of payloads to the lunar surface and even during its journey to the Moon. Bezos envisioned building a city on the Moon complete with robots.[34] He told NASA that the Blue Origin space venture could make Amazon-like deliveries to the Moon as part of his *Blue Moon* project.[35]

SpaceX is an American aerospace manufacturer and space transportation service company founded by Elon Musk in 2002.[36] It is also the first company that uses privately funded liquid–propellant rocket *Falcon 1* to reach the orbit.[37] *Falcon Heavy* is the most powerful rocket today.[38] It has a high liftoff thrust equal to approximately eighteen 747 aircraft at full power. It can deliver large payloads to orbits, only second to the Saturn V Moon rocket last flown in 1973.[39] *SpaceX Dragon* is a free-flying spacecraft designed to deliver both cargo and people to orbiting destinations. It is the only spacecraft that can deliver significant amounts of cargo. In fact, the *Dragon* spacecraft successfully docked with the space station ahead of schedule at 6:02 a.m. ET on March 3, 2019, becoming the first American spacecraft in history to autonomously dock with the ISS.[40] SpaceX *Starship* is a fully reusable transportation system, designed to carry both crew and payloads to the Earth orbit, the ISS, the Moon and Mars, and beyond. It will be an interplanetary transportation for human spaceflight development as well as building bases in other planets in the future.

Moon Village Project

Science fiction of lunar construction tells us a lot about how a Moon city would look like. But actually building a base on Moon is going to be a completely different game in reality. The immediate problem we will face is the availability of oxygen, food, water and electricity. These are factors limiting how long astronauts can stay.

Since 1960, scientists have conjectured that water ice could survive at the Moon's poles. NASA's Moon Mineralogy Mapper had images that evidenced the distribution of surface ice at the Moon's South Pole (left) and North Pole (right). That means much of this water ice could be accessible as a resource for future expeditions to explore. It could even be purified for drinking purposes.[41]

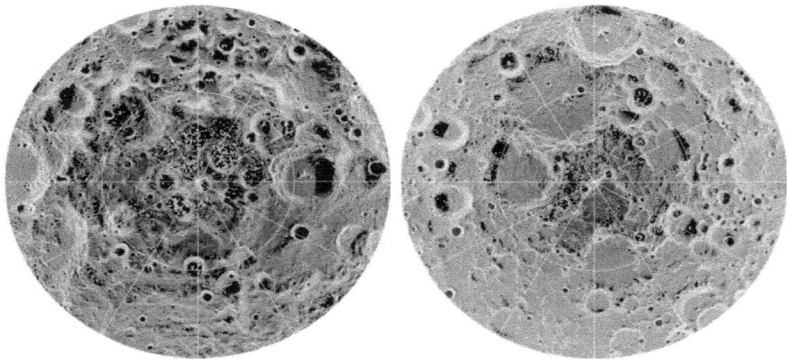

Figure 17.3:Distribution of surface ice at the Moon's south pole (left) and north pole
(right)

Source: NASA

With water, it is possible to get an endless supply of oxygen through electrolysis.

But where do we get electricity on the Moon?

The answer will be solar panels.

The lunar North Pole region exhibits a low quantity of sunlight. But the lunar South Pole features a region with crater rims exposed to near-constant solar illumination. This is a plausible destination for the novel Moon Village.

The next problem we will face is building the Moon Village. We need bulldozers and a lot of heavy equipment. We need to be able to forge metals to make things. If everything came from Earth for resupply, it would be very costly. World governments must look for alternate ways to develop a presence in space without bankrupting the planet. This is where 3D printers, nanotechnology, self-repairing robots and AI technology come in.

In the previous chapter, we looked at the miracle material of the 21st century - graphene. It is a superconductor and practically unbreakable. Its versatility makes it the perfect material to be used to build a Moon Village. So, the challenge in this century will be to mass-produce graphene. This would drastically reduce the cost of building infrastructure in space.

At the beginning of terraforming Moon, labor will be scarce. So, the key to constructing the novel Moon Village will require intelligent autonomous robots able to work together. Some of these robots will perform the task of exploring terrain otherwise hazardous or lethal for astronauts. Futuristic 3D printers will print the desired robots or parts to assemble buildings. Intelligent bulldozers and cranes will work together to transport those parts and assemble them into physical buildings and transportation systems. They will be able to learn and improve efficiency. A central AI will oversee the entire process and feedback to the command center on Earth.

Although the idea of the Moon village is pure science fiction, Europe, NASA, and even China are all currently working toward the goal of establishing a long-term presence on the Moon. In fact, the European Space Agency (ESA) and NASA are secretly currently planning a series of 37 rocket launches, both robotic and crew, to build a lunar settlement.[42]

But no matter who gets the credit in building a Moon Village, this project is going to be the greatest international collaboration in the history of mankind.

Making Life Multiplanetary

Earth is overpopulated, natural resources on Earth are depleting, and natural disasters are becoming severe due to climate change. Humanity is at a crossroad in history. We either stay on Earth forever and wait for some catastrophic event on Earth, or we become a multiplanetary species. I believe the future of humanity is in space. That is why making life multiplanetary is an important mission in the 21st century.

But which planet is suitable for humans?

Mercury is too close to the sun. Venus has super high pressure. Jupiter and Saturn have no solid surface to land. This leaves us only two options: Mars and the Moon.

If you have watched the 2015 Movie The Martian, you probably remember the challenge faced by the astronaut played by Matt Damon. The movie gave us a realistic taste of the difficulties colonists would face terraforming Mars. The ultra-low pressure in Mars, about 100 times thinner than Earth, means your blood would boil even at ambient temperature. The air on Mars is composed of more than 95% carbon dioxide. And worst of all, the red planet cannot deflect harmful radiation coming from space.[43]

Like the Moon, the red planet has got ice right in its soil. It has 5 million cubic kilometers of ice and 25 trillion metrics tons of carbon dioxide. NSAS's In situ resource utilization (ISRU) system will demonstrate technologies to use the Moon's resources to produce water, fuel, and other supplies, as well as leverage robotic capabilities to excavate and construct structures on the Moon. It will be converted into oxygen through electrolysis and then liquefied for storage. Carbon dioxide will be collected from the atmosphere to go through a process called *Sabatier* to produce methane for liquefaction storage.[44]

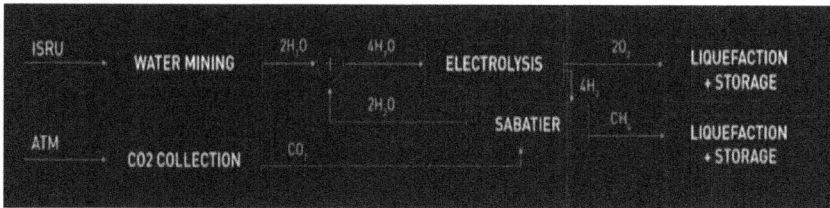

Figure 17.4: Water Mining and carbon dioxide collection on Mars

Source: NASA

But what about problems such as low pressure and low-temperature atmosphere on Mars that prohibit inhabitants?

Elon Musk proposed an idea to make Mars more Earth-like by dropping thermonuclear bombs on the ice at its poles. Doing so will melt the ice, and the greenhouse gas trapped inside will be released to heat up the atmosphere. His idea is to create self-sustaining warming to increase the temperature and pressure and make Mars habitable.[45]

To have a permanent presence on Mars, humanity's first agenda would be to build a permanent base for astronauts. At the beginning, it may be only a few volunteers who are willing to stay for a few months. Next, the population will increase, and they will begin exploiting raw material from the planet and set up solar power plants to harvest energy from the sun. With a way to heat the atmosphere, like Musk proposed, ice water will turn liquid. Once liquid water starts to flow, and with an abundance of carbon dioxide, agriculture will become possible. This is the vision of Robert Zubrin, an American aerospace engineer, the best known advocate of human exploration of Mars. This is how a new branch of human civilization will possibly look like on Mars in our future.

Chapter 18
Conclusion

Earth is a spaceship drifting in space, with our sun our energy supplier.

We are all astronauts inside it.

What makes this spaceship so special is that it keeps life regenerating onboard.

But, like any spaceship, if we don't look after it, it will cease to function.

I firmly believe that making Earth's finite resources sustainable for a human population heading towards nine billion, without deteriorating our ecosystem, is a huge challenge in the 21st century. As passengers of Spaceship Earth, people from different nationalities, races, colors, cultures must work together to address this challenge with our mind and technology.

Before closing, I wish to thank you for taking the time to read this book.

I hope *Crystal Balls of the 21st Century* has given you an overview of what will happen in our future. More importantly, I hope you all enjoy reading it.

I truly believe that humanity is at a crossroads in history, where the next twenty years will be completely unlike the last twenty years. We will be facing challenges in the 21st century, never seen before in the world. But right now, we must act. With technology and our genius, I am also confident that we can overcome those challenges and come out with a world filled with unprecedented prosperity.

After all, there is nothing in a caterpillar that tells you it is going to be a butterfly.

Introduction

1. Data source: Wiki, Interactive online World3 simulation, https://insightmaker.com/insight/1954/The-World3-Model-A-Detailed-World-Forecaster

2. Meadows, Donella H; Meadows, Dennis L; Randers, Jørgen; Behrens III, William W (1972). The Limits to Growth; A Report for the Club of Rome's Project on the Predicament of Mankind. New York: Universe Books. ISBN 0876631650. Retrieved 26 November 2017.

3. This information comes from McGill, the School of Computer Science, https://www.cs.mcgill.ca/~rwest/wikispeedia/wpcd/wp/b/Buckminster_Fuller.htm

4. This information comes from TIME Magazine, 10th January, 1964, http://content.time.com/time/covers/0,16641,19640110,00.html

5. Martin Bryant, "*20 years ago today, the World Wide Web opened to the public*", The Next Web, Aug 6, 2011, https://thenextweb.com/insider/2011/08/06/20-years-ago-today-the-world-wide-web-opened-to-the-public/

6. This information comes from Buckminster Fuller Institute, Tuesday, 13 January 1981, https://www.bfi.org/publications/fuller-shows-how-do-more-less

7. This information comes from Buckminster Fuller Institute, https://www.bfi.org/about-fuller/big-ideas/spaceshipearth

8. This information comes from Buckminster Fuller Institute, https://www.bfi.org/about-fuller/big-ideas/dymaxion-world/dymaxion-map

9. Martin Ford, "*Rise of the Robots*' and the threat of a jobless future", Basic Books, 2015,

10. Gordon E. Moore, "*Cramming more components onto integrated Circuits*", Intel Newsroom, https://newsroom.intel.com/wp-content/uploads/sites/11/2018/05/moores-law-electronics.pdf

11. This information comes from the University of Missouri – St Louis, https://www.umsl.edu/~siegelj/information_theory/projects/Bajramovic/www.umsl.edu/_abdcf/Cs4890/link1.html

12.   This information comes from Intel website
https://community.intel.com/t5/Processors/Transistor-count-of-Core-i7-2nd-generation-quot-Sandy-Bridge/td-p/353015

13.   Ray Kurzweil, "*The Law of Accelerating Returns*", kurzweilai.net, 7 March 2011, https://www.kurzweilai.net/the-law-of-accelerating-returns

14.   Priya Ganapati, "*Reverse-Engineering of Human Brain Likely by 2030, Expert Predicts*", Wired.com, 16 August 2010,
https://www.wired.com/2010/08/reverse-engineering-brain-kurzweil/

15.   This information comes from the National Human Genome Research Institute
https://www.genome.gov/human-genome-project/What

16.   Max Roser, Esteban Ortiz-Ospina and Hannah Ritchie, "*Life Expectancy*", Our World in Data, First published in 2013; last revised in October 2019,
https://ourworldindata.org/life-expectancy

17.   This information comes from Havard Stem Cell Institute
https://hsci.harvard.edu/new-steps-forward-cell-reprogramming

18.   Kathiann Kowalski, "*Silencing genes — to understand them*", Science News for Students, 27 March, 2015, https://www.sciencenewsforstudents.org/article/silencing-genes-understand-them

Chapter 1

1.   Max Roser, Esteban Ortiz-Ospina and Hannah Ritchie, "*World Population*", Our World in Data, First published in 2013; last revised in October 2019.
https://ourworldindata.org/world-population-growth

2.   Natalie Wolchover, "*How Many People can Earth Support?*", Live Science, 11 October 2011, https://www.livescience.com/16493-people-planet-earth-support.html

3.   Klein, D.R. 1968. "*The introduction, increase, and crash of reindeer on St. Matthew Island*", J. Wildlife Management 32: 350-367.

4.   This information comes from United Nations, https://www.un.org/en/sections/issues-depth/population/index.html

5. Max Roser, "*Fertility Rate*", Our World in Data, First published in 2014; substantive revision published on December 2, 2017. https://ourworldindata.org/fertility-rate

6. This information comes from the Pew Research Center, https://www.pewresearch.org/global/2014/01/30/chapter-4-population-change-in-the-u-s-and-the-world-from-1950-to-2050/

7. This information comes from the PeenState College of Earth and Mineral Sciences https://www.e-education.psu.edu/geog30/node/328

8. This information comes from United Nations, https://www.un.org/development/desa/en/news/population/world-population-prospects-2017.html

9. Adam Morton, Nick Evershed and Graham Readfearn, "*Australia bushfires factcheck: are this year's fires unprecedented?*", the Guardian, 23 November 2019 https://www.theguardian.com/australia-news/2019/nov/22/australia-bushfires-factcheck-are-this-years-fires-unprecedented

10. Jon Henley and Angela Giuffrida, "*Two people die as Venice floods at highest level in 50 years*", the Guardian, 23 November 2019 https://www.theguardian.com/environment/2019/nov/13/waves-in-st-marks-square-as-venice-flooded-highest-tide-in-50-years

Chapter 2

1. P. Sreevani, "*Wood as a renewable source of energy and future fuel*", AIP Conference Proceedings, 3 August 2018 https://aip.scitation.org/doi/10.1063/1.5047972

2. Mary Bellis, "*The History of the Water Wheel*", ThoughtCo., 24 November 2019, https://www.thoughtco.com/history-of-waterwheel-4077881

3. This information comes from the U.S. Energy Information Administration, "*Wind Explained: History of wind power*", https://www.eia.gov/energyexplained/wind/history-of-wind-power.php

4.   John U. Nef, "*As Early Energy Crisis and its Consequences*", https://nature. berkeley.edu/er100/readings/Nef_1977.pdf

5.   The information comes from Kentucky Foundation, "*Who discovered coal and who discovered that it could be used as a heat source and when?*", http://www. coaleducation.org/q&a/who_discovered_coal.htm

6.   This information comes from bbc.co.uk, "*Newcomen's steam revolution*", http:// www.bbc.co.uk/devon/discovering/famous/thomas_newcomen.shtml

7.   This information comes from bbc.co.uk, "*James Watt (1736 – 1819)*", http:// www.bbc.co.uk/history/historic_figures/watt_james.shtml

8.   Sabrina Son, This information comes from Tiny Pulse, "*The Origin of the 8 Hour Work Day — and Why It No Longer Matters*",16 February 2017 https://www.tinypulse.com/blog/the-origin-of-the-8-hour-work-day-and-why-it-no-longer-matters

9.   Jennifer Latson, "*How the American Oil Industry Got Its Start*", TIME, 27 August 2015, https://time.com/4008544/american-oil-well-history/

10.   This information comes from the New Bedford Whaling Museum, "*Whales and Hunting*", https://www.whalingmuseum.org/learn/research-topics/overview-of-north-american-whaling/whales-hunting

11.   This information comes from the Environmental Literacy Council, "*Petroleum History*", https://enviroliteracy.org/energy/fossil-fuels/petroleum-history/

12.   This information comes from the Constitutional Rights Foundation, "*BRIA 16 2 b Rockefeller and the Standard Oil Monopoly*", https://www.crf-usa.org/bill-of-rights-in-action/bria-16-2-b-rockefeller-and-the-standard-oil-monopoly.html

13.   This information comes from the Constitutional Rights Foundation, "*BRIA 16 2 b Rockefeller and the Standard Oil Monopoly*", https://www.crf-usa.org/bill-of-rights-in-action/bria-16-2-b-rockefeller-and-the-standard-oil-monopoly.html

14.   This information comes from the Science History Institute, "*History and Future of Plastic*", https://www.sciencehistory.org/the-history-and-future-of-plastics

15.   Brian C. Black, "*How World War I ushered in the century of oil*", The

Conversation, 4 April 2017, https://theconversation.com/how-world-war-i-ushered-in-the-century-of-oil-74585

16.   This information comes from NaturalGas.org, http://naturalgas.org/overview/history/

17.   This information comes from the U.S. Energy Information Administration https://www.eia.gov/energyexplained/natural-gas/natural-gas-pipelines.php

18.   This information comes from BBC, "*Benjamin Franklin (1706 – 1790)*", http://www.bbc.co.uk/history/historic_figures/franklin_benjamin.shtml

19.   This information comes from the Franklin Institute, "*Benjamin Franklin Experiment*", https://www.fi.edu/benjamin-franklin/kite-key-experiment

20.   This information comes from BBC, "*Michael Faraday (1791 – 1867)*", http://www.bbc.co.uk/history/historic_figures/faraday_michael.shtml

21.   This information is from the McGraw Commons, "*Dynamo*", http://commons.princeton.edu/motorcycledesign/wp-content/uploads/sites/70/2018/06/Dynamo.pdf

22.   This information comes from Edison Tech Centre, "*The History of Electrification*", http://edisontechcenter.org/HistElectPowTrans.html

23.   This information comes from energy.gov, "*The War of the Current*", 18 November 2014, https://www.energy.gov/articles/war-currents-ac-vs-dc-power

24.   Gilbert King, "*Edison vs. Westinghouse: A Shocking Rivalry*", Smithsonian Magazine, 11 October 2011, https://www.smithsonianmag.com/history/edison-vs-westinghouse-a-shocking-rivalry-102146036/,

25.   Tony Long, "*Jan. 4, 1903: Edison Fries an Elephant to Prove His Point*", 1 April 2008, https://www.wired.com/2008/01/dayintech-0104/

26.   Elizabeth Nix, "*How Edison, Tesla and Westinghouse Battled to Electrify America*", 24 October 2019, https://www.history.com/news/what-was-the-war-of-the-currents

27.   This information comes from the International Institute for Sustainable Development, "*Renewable Energy Investment to Surpass USD 2.5 Trillion for 2010-2019, UNEP Report Finds*", 10 September 2019,

https://sdg.iisd.org/news/renewable-energy-investment-to-surpass-usd-2-5-trillion-for-2010-2019-unep-report-finds/

28.   Tim Radford, "*The World Is Not Converting to Renewable Energy Fast Enough to Save It*", 28 March 2016, https://www.truthdig.com/articles/the-world-is-not-converting-to-renewable-energy-fast-enough-to-save-it/

29.   This information comes from the Constitutional Rights Foundation, "*BRIA 16 2 b Rockefeller and the Standard Oil Monopoly*", https://www.crf-usa.org/bill-of-rights-in-action/bria-16-2-b-rockefeller-and-the-standard-oil-monopoly.html

30.   Anthony Sampson, "*The Seven Sisters: The Great Oil Companies and the World They Shaped*", Energy Today, https://www.energytoday.net/conventional-energy/the-seven-sisters-the-great-oil-companies-and-the-world-they-shaped/

31.   This information comes from Focus Economics, "*The History of OPEC: Has it been a Success?*", https://www.focus-economics.com/blog/opec-history-has-opec-been-a-success

32.   This information comes from the Organization of the Petroleum Exporting Countries, https://www.opec.org/opec_web/en/

33.   Data Source: U.S. Energy Information Administration, "*What countries are the top producers and consumers of oil?*" , 1 April 2020, https://www.eia.gov/tools/faqs/faq.php?id=709&t=6

34.   Data Source: U.S. Energy Information Administration, "*What countries are the top producers and consumers of oil?*" , 1 April 2020, https://www.eia.gov/tools/faqs/faq.php?id=709&t=6

35.   Data Source: Worldometer, "*Oil left in the world*", https://www.worldometers.info/oil/

36.   Data Source: Worldometer, "*Oil left in the world*", https://www.worldometers.info/oil/

37.   Data Source: Central Intelligence Agency, "*Country comparison – Crude Oil – proved reserves*", https://howmuch.net/articles/worlds-biggest-crude-oil-reserves-by-country

38. Data Source: Central Intelligence Agency, "*Country comparison – Crude Oil – proved reserves*", https://howmuch.net/articles/worlds-biggest-crude-oil-reserves-by-country

39. Torren Peebles, "*Development of Hubbert's Peak Oil Theory and Analysis of its Continued Validity for U.S. Crude Oil Production*", 5 May 2017, https://earth.yale.edu/sites/default/files/files/Peebles_Senior_Essay.pdf

40. This information comes from HistoryofInformation.com, "*The First Successful Oil Well is Drilled in Titusville, Pennsylvania*", https://www.historyofinformation.com/detail.php?id=2650

41. Marie Plummer Minniear, originally published by The Oil Drum,"*Forecasting the permanent decline in Global Petroleum production*", 11 December 2009, https://www.resilience.org/stories/2009-12-11/forecasting-permanent-decline-global-petroleum-production/

42. The information comes from Parliament of Australia, "*Chapter 3 - 'Peak oil' concerns about future oil supply*", https://www.aph.gov.au/Parliamentary_Business/Committees/Senate/Rural_and_Regional_Affairs_and_Transport/Completed_inquiries/2004-07/oil_supply/report/c03

43. Euan Mearns, "*ERoEI for Begineers*", Euanmearns.com, 25 May 2016, http://euanmearns.com/eroei-for-beginners/

44. Marc Lallanilla, "*Facts about Fracking*", Live Science, 10 February 2018, https://www.livescience.com/34464-what-is-fracking.html

45. Catherine Lane, "*Are lithium ion solar batteries the best energy storage option?*", Solar Reviews, 29 July 2020, https://www.solarreviews.com/blog/are-lithium-Ion-the-best-solar-batteries-for-energy-storage

46. This information comes from Intertek, "*The Future of Battery Technologies – Part V Environmental Considerations for Lithium Batteries*", https://www.intertek.com/uploadedFiles/Intertek/Divisions/Commercial_and_Electrical/Media/PDF/Battery/Environmental-Considerations-for-Lithium-Batteries-White-Paper.pdf

47. Paul Breeze, "*Power System Energy Storage Technologies*", Science Direct,

2018, https://www.sciencedirect.com/topics/engineering/hydrogen-energy-storage

48.   This information comes from the U.S. Department of Energy, "*Hydrogen Production and Distribution*", https://afdc.energy.gov/fuels/hydrogen_production.html

49.   Vladimir Strezov, "*Explainer: what is hydroelectricity?*", The Conversation, 16 April 2013, https://theconversation.com/explainer-what-is-hydroelectricity-12931

50.   Duncan Graham-Rowe, "*Hydroelectric power's dirty secret revealed*", News Scientists, 24 February 2005, https://www.newscientist.com/article/dn7046-hydroelectric-powers-dirty-secret-revealed/

51.   This information comes from CarbonBrief, "*Energy return on investment – which fuels win?*", 20 March 2013, https://www.carbonbrief.org/energy-return-on-investment-which-fuels-win

52.   This information comes from the International hydropower Association, "*China*", May 2019, https://www.hydropower.org/country-profiles/china

53.   This information comes from USGS, "*Three Gorges Dam: The World's Largest Hydroelectric Plant*", https://www.usgs.gov/special-topic/water-science-school/science/three-gorges-dam-worlds-largest-hydroelectric-plant?qt-science_center_objects=0#qt-science_center_objects

54.   Nicholas Withers, "*Building the world's biggest dam*", NES Fircroft, 28 September 2020, https://www.fircroft.com/blogs/building-the-worlds-biggest-dam-02722815435

55.   Charles Rotter, "*Wind farm turbines wear sooner than expected, says study*", WUWT (Watts Up With That), 29 December 2018, https://wattsupwiththat.com/2018/12/29/wind-farm-turbines-wear-sooner-than-expected-says-study/

56.   This information comes from the Australian Renewable Energy Agency, "*Ocean Energy*", 18 August 2020, https://arena.gov.au/renewable-energy/ocean/

57.   This information comes from the U.S. Energy Information Administration, "*Hydropower explained - Ocean thermal energy conversion*", Last updated: December 4, 2019, https://www.eia.gov/energyexplained/hydropower/ocean-thermal-energy-conversion.php

58.   This information comes from the International Renewable Energy Agency, "*Ocean Thermal Energy Conversion*", June 2014, https://www.irena.org/documentdownloads/publications/ocean_thermal_energy_v4_web.pdf

59.   Zachary Shahan, "*Electric Motors Use 45% of Global Electricity, Europe Responding*", CleanTechnica, 16 June 2011, https://cleantechnica.com/2011/06/16/electric-motors-consume-45-of-global-electricity-europe-responding-electric-motor-efficiency-infographic/

60.   This information comes from the International Electrotechnical Commission, "*Examples by industry sector Electric motors*", https://www.iec.ch/perspectives/government/sectors/electric_motors.htm

61.   This information comes from the Department of Industry, Science, energy and Resources in the Australian Government, "*Motors and variable speed drives*", https://www.energy.gov.au/business/equipment-and-technology-guides/motors-and-variable-speed-drives

62.   Zhenya Liu, "*Global Energy Interconnection*", ScienceDirect, 2016, https://www.sciencedirect.com/book/9780128044056/global-energy-interconnection

63.   Jim Cahill, "*Ultra-High Voltage Transmission (UHV) – A New Way to Move Power*", Emerson, 9 January 2015, https://www.emersonautomationexperts.com/2015/industry/power-generation/ultra-high-voltage-transmission-uhv-a-new-way-to-move-power/

64.   Liza Reed, M. Granger Morgan, Parth Vaishnav, Daniel Erian Armanios, "*Converting existing transmission corridors to HVDC is an overlooked option for increasing transmission capacity*", 9 July 2019, https://www.pnas.org/content/116/28/13879

65.   The information comes from Green Car Congress, "*First 800-kV High-Voltage Direct-Current Link in China Now Fully Operational; Transmission of Hydro Power to Pearl River Delta*", 23 June 2010, https://www.greencarcongress.com/2010/06/hvdc-20100623.html

66.   The information comes from Azom.com, "*High Temperature Superconducting*"

Transmission Cables", 11 Oct 2001, https://www.azom.com/article.aspx?ArticleID=942

67.   The information comes from IEC whitepaper, *"Global Energy Interconnection"*, http://www.terrawatts.com/GEI-whitepaper.pdf

68.   By Edmund Downie, *"China's Vision for a Global - The Politics of Global Energy Interconnection Grid"*, 13 February 2019, https://reconnectingasia.csis.org/analysis/entries/global-energy-interconnection/

69.   Quang-Dung Ho, Tho Le-Ngoc, *"Smart Grid Communications Networks: Wireless Technologies, Protocols, Issues, and Standards1"*, Science Direct, 2013, https://www.sciencedirect.com/topics/engineering/phasor-measurement-units

70.   This information comes from GE, *"Wide Area Monitoring System (WAMS): PhasorPoint Applications for advanced energy management"*, https://www.ge.com/digital/sites/default/files/download_assets/wams-phasorpoint-applications-from-ge-digital.pdf

71.   B. Panajotovic, M. Jankovic and B. Odadzic, *"ICT and smart grid,"* 2011 10th International Conference on Telecommunication in Modern Satellite Cable and Broadcasting Services (TELSIKS), Nis, 2011, pp. 118-121, doi: 10.1109/TELSKS.2011.6112018.

Chapter 3

1.   This information comes from the United Nations Climate Change, *"What is the Paris Agreement?"*, https://unfccc.int/process-and-meetings/the-paris-agreement/what-is-the-paris-agreement

2.   Melissa Denchak, *"Paris Climate Agreement: Everything You Need to Know"*, NRDC, 12 December 2018, https://www.nrdc.org/stories/paris-climate-agreement-everything-you-need-know

3.   This information comes from IPCC, *"Special Report: Global Warming of 1.5 ºC"*, https://www.ipcc.ch/sr15/chapter/spm/

4.   This information comes from NASA Earth observatory, https://earthobservatory.nasa.gov/world-of-change/global-temperatures

5.	This information comes from NASA Earth observatory, https://earthobservatory.nasa.gov/world-of-change/global-temperatures

6.	This information comes from the National Snow & Ice Data Center, "*Climate Change in the Arctic*", https://nsidc.org/cryosphere/arctic-meteorology/climate_change.html

7.	This information comes from Science Daily, "*Could we cool Earth with an ice-free Arctic?*", 10 December 2019, https://www.sciencedaily.com/releases/2019/12/191210111641.htm

8.	This information comes from Met Office, "*What is the Atlantic Meridional Overturning Circulation?*", https://www.metoffice.gov.uk/weather/learn-about/weather/oceans/amoc

9.	This information comes from the National Ocean Service, "*What is the global ocean conveyor belt?*", 25 June 2018, https://oceanservice.noaa.gov/facts/conveyor.html

10.	This information comes from WWF, "*Why are glaciers and sea ice melting?*", https://www.worldwildlife.org/pages/why-are-glaciers-and-sea-ice-melting

11.	Jon Henley, Angela Giuffrida, "*Two people die as Venice floods at highest level in 50 years*", The Guardian, 14 November 2019, https://www.theguardian.com/environment/2019/nov/13/waves-in-st-marks-square-as-venice-flooded-highest-tide-in-50-years

12.	Gelu Sulugiuc, "*Sea level rise underestimated: scientists*", ABC Science, 11 March 2009, https://www.abc.net.au/science/articles/2009/03/11/2513003.htm

13.	This information comes from the National Ocean Service, "*Is Sea level rising?*", last updated 10 September 2019, https://oceanservice.noaa.gov/facts/sealevel.html

14.	This information comes from the National Snow & Ice Data Center, https://nsidc.org/cryosphere/quickfacts/icesheets.html

15.	This information comes from the National Snow & Ice Data Center, https://nsidc.org/cryosphere/quickfacts/icesheets.html

16.	Kelly McLaughlin, "*More than a billion animals are feared dead in Australia's bushfires*", Business Insider, 8 January 2020, https://www.businessinsider.com/australia-bushfires-one-billion-animals-feared-dead-2020-1?r=AU&IR=T

17.   This information comes from the World Weather Attribution, "*Attribution of the Australian bushfire risk to anthropogenic climate change*", 10 January 2020, https://www.worldweatherattribution.org/bushfires-in-australia-2019-2020/

18.   This information comes from BBC.com, "*Australia to cull thousands of camels*", 8 January 2020, https://www.bbc.com/news/newsbeat-51032145

19.   This information comes from the Bureau of Meteorology from the Australian Government, "*Annual Climate Statement 2018*", http://www.bom.gov.au/climate/current/annual/aus/2018/

20.   Hannah Ritchie and Max Roser, "*Plastic Pollution*", Our World in Data, September 2018, https://ourworldindata.org/plastic-pollution

21.   Brooke Bauman, "*How plastics contribute to climate change*", Yale Climate Connections, 20 August 2019, https://www.yaleclimateconnections.org/2019/08/how-plastics-contribute-to-climate-change/

22.   Lorin Hancock, "*10 worst single-use plastics and eco-friendly alternatives*", WWF, 1 July 2020, https://www.wwf.org.au/news/blogs/10-worst-single-use-plastics-and-eco-friendly-alternatives

23.   This information comes from Surfers Against Sewage, "*Plastic Pollution – Facts and Figures*", https://www.sas.org.uk/our-work/plastic-pollution/plastic-pollution-facts-figures/

24.   This information comes from ABC news, "*More plastic than fish in the oceans by 2050, report warns*", 21 January 2016, https://www.abc.net.au/news/2016-01-21/more-plastic-than-fish-in-the-oceans-by-2050-report-warns/7105936

25.   This information comes from the Bureau of Reclamation, "*Water Facts – Worldwide water supply*", https://www.usbr.gov/mp/arwec/water-facts-ww-water-sup.html

26.   This information comes from the Life You can Save Australia, https://www.thelifeyoucansave.org.au/causes-to-support/water-charities/?gclid=Cj0KCQjwwuD

7BRDBARIsAK_5YhVNPLs6tN-A8cOU-mSrrtiqSPO4tdPm5BaCfMZaSYrG9p-dJ2CP7BEaAg3_EALw_wcB

27.   This information comes from CUESA, "*10 Ways Farmers Are Saving Water*", 15 August 2014, https://cuesa.org/article/10-ways-farmers-are-saving-water

28.   This information comes form the UN environment programme, "*Global Environment Outlook*", https://www.unenvironment.org/global-environment-outlook

Chapter 4

1.   This information comes from the Federal Reserve history, "*Nixon Ends Convertibility of US Dollars to Gold and Announces Wage/Price Controls*", August 1971, https://www.federalreservehistory.org/essays/gold_convertibility_ends

2.   This information comes from the World Gold Council, "*The classical Gold Standard*", https://www.gold.org/about-gold/history-of-gold/the-gold-standard

3.   This information comes from the International Monetary Fund, https://www.imf.org/external/index.htm

4.   This information comes from the World Bank, https://www.worldbank.org/en/who-we-are/ibrd

5.   Michael Bordo, "*The operation and demise of the Bretton Woods system: 1958 to 1971*", 23 April 2017, https://voxeu.org/article/operation-and-demise-bretton-woods-system

6.   Alexander Chipman Koty and Qian Zhou, "*A Guide to Minimum Wages in China in 2020*", 29 April 2020, https://www.china-briefing.com/news/minimum-wages-china-2020/

Chapter 5

1.   Ray Kurzwell, "*The Law of Accelerating Returns*", kurzweil, 7 March 2001, https://www.kurzweilai.net/the-law-of-accelerating-returns

2.   This information comes from the University of Missouri-St Louis https://www.umsl.edu/~siegelj/information_theory/projects/Bajramovic/www.umsl.

edu/_abdcf/Cs4890/link1.html

3.   Joel Hruska, "*We've Never Seen Intel Struggle Like This*", ExtremeTech, 27 July 2020, https://www.extremetech.com/computing/313208-weve-never-seen-intel-struggle-like-this

4.   Jon Porter, "*Google confirms quantum supremacy breakthrough*", The Verge, 23 October 2019, https://www.theverge.com/2019/10/23/20928294/google-quantum-supremacy-sycamore-computer-qubit-milestone

5.   Elizabeth Gibney, "*Hello quantum world! Google publishes landmark quantum supremacy claim*", Nature, 23 October 2019, https://www.nature.com/articles/d41586-019-03213-z

6.   Alessandro Rossi and M. Fernando Gonzalez-Zalba, "*Neven's Law: why it might be too soon for a Moore's Law for quantum computers*", the conversation, 25 July 2019, https://theconversation.com/nevens-law-why-it-might-be-too-soon-for-a-moores-law-for-quantum-computers-120706

7.   Jo Best, "*What is a brain-computer interface? Everything you need to know about BCIs, neural interfaces and the future of mind-reading computers*", ZDNet, 13 November 2019, https://www.zdnet.com/article/what-is-bci-everything-you-need-to-know-about-brain-computer-interfaces-and-the-future-of-mind-reading-computers/

8.   Luke Walton, "*Cells programmed like computers to fight disease*", phy.org, 18 September 2017, https://phys.org/news/2017-09-cells-disease.html

Chapter 6

1.   Andrew Lycett, "*Breaking Germany's Enigma Code*", BBC.co.uk, 17 February 2011, http://www.bbc.co.uk/history/worldwars/wwtwo/enigma_01.shtml

2.   Macgregor Campbell, "*Unbreakable: The second world war's last Enigma*", New Scientist, 18 My 2011, https://www.newscientist.com/article/mg21028134-100-unbreakable-the-second-world-wars-last-enigma/

3.   A. M. Turing; On Computable Numbers, with an Application to the Entscheidungsproblem, Proceedings of the London Mathematical Society, Volume s2-42

Issue 1, 1937,

4. Miss Cellania, "*Ada Lovelace: The First Computer Programmer*", Mental Foloss, 13 October 2015, https://www.mentalfloss.com/article/53131/ada-lovelace-first-computer-programmer

5. Suw, "*Explore Ada Lovelace's Bernoulli program with Wolfram*", findingada. com, 9 October 2016, https://findingada.com/blog/2016/10/09/explore-ada-lovelaces-bernoulli-program-with-wolfram/

6. The information comes from computerhistory.org, "*The Engines*", https://www. computerhistory.org/babbage/engines/

7. The information comes from computerhistory.org, "*The Engines*", https://www. computerhistory.org/babbage/engines/

8. The information comes from adacore.com, "*Celebrating Ada Lovelace: The Pioneer of Programming*", https://www.adacore.com/press/celebrating-ada-lovelace-the-pioneer-of-programming

9. Sebastian Anthony, "*The history of computer storage*", ExtremeTech, 3 August 2011, https://www.extremetech.com/computing/90156-the-history-of-computer-storage-slideshow/2

10. The information comes from "*Walt at Random*", https://walt.lishost. org/2011/12/when-grandpa-used-80-column-cards/

11. Roy, Gautam (2007). Computer Studies for Engineering Students. Mumbai, IN: Allied Publishers Limited. p. 10. ISBN 9788184242119. Retrieved July 28, 2016.

12. The information comes from History-computer.com, "*SAGE*" https://history-computer.com/ModernComputer/Electronic/SAGE.html

13. This information comes from computer.org, "*J.Presper Eckert*", https://www.computer.org/profiles/presper-eckert

14. B. Jack Copeland, "*The Modern History of Computing*", The Stanford Encyclopedia of Philosophy, 18 December 2000, https://stanford.library.sydney.edu.au/entries/computing-history/

15. The information comes from The University of Manchester, "*The Mercury*

Acoustic Delay Line", http://curation.cs.manchester.ac.uk/computer50/www.computer50. org/mark1/notes.html

16.   The information comes from Computer History Museum, "*1953: Whirlwind Computer Debuts Core Memory*", https://www.computerhistory.org/storageengine/ whirlwind-computer-debuts-core-memory/

17.   The information comes from history.com, "*UNIVAC, the first commercially produced digital computer, is dedicated*", https://www.history.com/this-day-in-history/ univac-computer-dedicated

18.   Suzanne Deffree, "*IBM intros 1st computer disk storage unit, September 13, 1956*", EDN, 13 September 2019, https://www.edn.com/ibm-intros-1st-computer-disk-storage-unit-september-13-1956/

19.   Gary Brown, "*How Floppy Disk Drives Work*", How Stuff Works, 26 February 2001, https://computer.howstuffworks.com/floppy-disk-drive1.htm

20.   Andy Patrizio, "*IDC: Expect 175 zettabytes of data worldwide by 2025*", Networkworld, 3 December 2018, https://www.networkworld.com/article/3325397/idc-expect-175-zettabytes-of-data-worldwide-by-2025.html

21.   Desire Athow, World's largest hard drive coming "*within months*", Techradar, 5 September 2019, https://www.techradar.com/au/news/20tb-by-2020-worlds-largest-hard-drive-to-land-within-months

22.   Douglas Heavem "*Books and JavaScript stored in DNA molecules*", News Scientists, 16 August 2012, https://www.newscientist.com/article/dn22190-books-and-javascript-stored-in-dna-molecules/

23.   Chris Mellor, "*Catalog claims DNA data storage is economically feasible for the first time*", blocksandfiles.com, 18 March 2020, https://blocksandfiles. com/2020/03/18/catalog-cdna-data-storage-economically-feasible/

24.   The information comes from columbia.edu, "*John Vincent Atanasoff The father of the computer*", http://www.columbia.edu/~td2177/JVAtanasoff/JVAtanasoff. html

25.   Mary Bellis, "*The Atanasoff-Berry Computer: The First Electronic*

Computer", ThoughCo, 23 February 2019, https://www.thoughtco.com/john-atanasoff-and-clifford-berry-inventors-4078350

26.   The information comes from techoblasto.blogspot.com, "*Atanasoff - Berry Computer (ABC)*", https://techoblasto.blogspot.com/2018/11/introduction-atanasoff-berry-computer.html

27.   Sam Byford, "*Colossus: how the first programmable electronic computer saved countless lives*", the verge, 12 March 2012, https://www.theverge.com/2012/3/12/2864068/colossus-first-programmable-electronic-computer

28.   The information comes from 1806engineering.com, "*The Post Office Research Station and the Colossus code-breaking computer*", https://www.1806engineering.com/post/the-post-office-research-station-and-the-colossus-code-breaking-computer

29.   The information comes from theevolutionofcomputers.wordpress.com, "*First Generation Computers*", https://theevolutionofcomputers.wordpress.com/first-generation-computers/

30.   Samantha Bookman, "*William Shockley, John Bardeen & Walter Brattain, co-invented the transistor*", 4 October 2011, https://www.fiercetelecom.com/special-report/william-shockley-john-bardeen-walter-brattain-co-invented-transistor-0

31.   J.M.K.C. Donev et al. (2020). Energy Education - Dopant [Online]. Available: https://energyeducation.ca/encyclopedia/Dopant.

32.   The information comes from IBM, "*IBM 608 calculator*" https://www.ibm.com/ibm/history/exhibits/vintage/vintage_4506VV2214.html

33.   Carolyn E. Tajnai, "*Fred Terman, The father of silicon Valley*", Stanford University, May 1985, http://forum.stanford.edu/carolyn/terman

34.   The information comes from siliconvalleyhistorical.org, "*Entrepreneurs*", https://www.siliconvalleyhistorical.org/entrepreneurs

35.   Michael Riordan, "*The Lost History of the Transistor*", IEEE Spectrum, 30 April 2004, https://spectrum.ieee.org/tech-history/silicon-revolution/the-lost-history-of-the-transistor

36.   Philippe Lacomme, "*Point-Contact Transistor*", ScienceDirect, 2001, https://

www.sciencedirect.com/topics/engineering/point-contact-transistor

37.   Marcio S Galli, "*Openness, risk taking, design, and consciousness*", Medium, 29 March 2017, https://medium.com/@taboca/openness-risk-taking-design-and-consciousness-4966e7be505

38.   Mark Hall, "*Intel*", Britannica, https://www.britannica.com/topic/Intel.

39.   This information comes from the National Inventors Hall of Fame, "*Dawon Kahng – MOSFET*", https://www.invent.org/inductees/dawon-kahng

40.   This information comes from History-computer.com, "*Integrated Circuit*" https://history-computer.com/ModernComputer/Basis/IC.html

41.   This information comes from hypertextbook.com , "*Number of Transistors on an Integrated Circuit*", https://hypertextbook.com/facts/2001/SerafinaShishkova.shtml

42.   Sebastian Anthony, "*Intel 4004, the first CPU, is 40 years old today*", ExtremeTech, 15 November 2011,https://www.extremetech.com/computing/105029-intel-4004-the-first-cpu-is-40-years-old-today

43.   This information comes from TSMC (Taiwan Semiconductor Manufacturing Corporation), "*About TSMC*", https://www.tsmc.com/english/aboutTSMC/index.htm

44.   Juli Clover, "*Apple Silicon Arm Macs: Coming in Late 2020*", MacRumours, 9 September 2020, https://www.macrumors.com/guide/apple-silicon/

45.   Paul Alcorn, "*Intel's 7nm is Broken, Company Announces Delay Until 2022, 2023*", Tomshardware.com, 23 July 2020, https://www.tomshardware.com/news/intel-announces-delay-to-7nm-processors-now-one-year-behind-expectations

46.   Eric Conrad, Seth Misenar and Joshua Feldman, "*Complex Instruction Set Computer*", ScienceDirect, 2016, https://www.sciencedirect.com/topics/computer-science/complex-instruction-set-computer

47.   Andrew N. Sloss, Dominic Symes and Chris Wright, "*Reduced Instruction Set Computer*", ScienceDirect, 2004, https://www.sciencedirect.com/topics/computer-science/reduced-instruction-set-computer

48.   This information comes from ScienceDirect, "*Von Neumann Architecture*" https://www.sciencedirect.com/topics/computer-science/von-neumann-architecture

49.   The information comes from IBM, "*Celebrating Abu Sebastian on National Inventors'Day*", https://www.ibm.com/blogs/research/2018/02/abu-sebastian-on-national-inventors-day/

Chapter 7

1.   This information comes from history.com, "*The Invention of the Internet*," updated: 28 October 2019, Original: 30 July 2010, https://www.history.com/topics/inventions/invention-of-the-internet

2.   This information comes from nternethalloffame.org, "*J.C.R. Licklider*", https://www.internethalloffame.org/inductees/jcr-licklider

3.   J.C.R. Licklider and Robert W. Taylor, "*The Computer as a Communication Device*", Science and technology, 1968, https://signallake.com/innovation/LickliderApr68.pdf

4.   This information comes from the Rochester Institute of Technology, "*J. C. R. Licklider*", https://www.cs.rit.edu/~rpretc/imm/project1/biography.html

5.   Amy Akmal, "*Internet50 Press*", UCLA Samueli School of Engineering, https://samueli.ucla.edu/internet50-press/

6.   Giovanni Navarria, "*How the Internet was born: the ARPANET Comes to Life*", 2 November 2016, https://theconversation.com/how-the-internet-was-born-the-arpanet-comes-to-life-68062

7.   Barry M. Leiner, Vinton G. Cerf, David D. Clark, Robert E. Kahn, Leonard Kleinrock, Daniel C. Lynch, Jon Postel, Larry G. Roberts, Stephen Wolff, "*Brief History of the Internet*", Internet Soceity, 1997, https://www.internetsociety.org/internet/history-internet/brief-history-internet/

8.   Timothy B. Lee, "*40 maps that explain the Internet*", vox.com, 2 June 2014, https://www.vox.com/a/internet-maps

9.   The information comes from the University System of Georgia, "*A Brief History of the Internet*", https://www.usg.edu/galileo/skills/unit07/internet07_02.phtml

10.   This information comes from networkencyclopedia.com, "*Networking History*

1980", https://networkencyclopedia.com/networking-history-1980/

11.   Dr. Farid Farahmand, "*Message Switching*", Central Connecticut State University http://web.sonoma.edu/users/f/farahman/sonoma/courses/cet543/resources/ch70_message_swiching.pdf

12.   This information comes from www.w3.org, "*A history of HTML*", Addison Wesley Longman, 1998, https://www.w3.org/People/Raggett/book4/ch02.html

13.   This information comes from the World Wide Web Foundation, "*History of the Web*", https://webfoundation.org/about/vision/history-of-the-web/

14.   Martin Bryant, "*20 years ago today, the World Wide Web opened to the public*", The Next Web, 6 August 2011, https://thenextweb.com/insider/2011/08/06/20-years-ago-today-the-world-wide-web-opened-to-the-public/

15.   Steven J. Vaughan-Nichols, "*Mosaic turns 25: The beginning of the modern web*", zdnet.com, 15 April 2018, https://www.zdnet.com/article/mosaics-birthday-25-years-of-the-modern-web/

16.   Alex McPeak, "*A Brief History of Web Browsers and How They Work*", Cross Browser Testing, 24 January 2018, https://crossbrowsertesting.com/blog/test-automation/history-of-web-browsers/

17.   Kaleigh Moore, "*Ecommerce 101 + The History of Online Shopping: What The Past Says About Tomorrow's Retail Challenges*", Bigcommerce, https://www.bigcommerce.com/blog/ecommerce/

18.   Joe Miller, "*Jonathon Fletcher: forgotten father of the search engine*", BBC News, 3 September 2013, https://www.bbc.co.uk/news/technology-23945326

19.   John Battelle, "*The Birth of Google*", Wired, 8 January 2005, https://www.wired.com/2005/08/battelle/

20.   Andy Patrizio, "*Opinion ICQ, the original instant messenger, turns 20*", Networkworld, 18 November 2016, https://www.networkworld.com/article/3142451/icq-the-original-instant-messenger-turns-20.html

21.   Josh Constine, "*AOL Instant Messenger is shutting down after 20 years*", Techcrunch, 7 October 2017,

https://techcrunch.com/2017/10/06/aol-instant-messenger-shut-down/

22.   The information comes from China IT News, "*From OICQ to instant messaging is no longer pure*", 7 October 2019, https://www.firstxw.com/view/239752.html

23.   This information comes from Britannica.com , https://www.britannica.com/technology/Skype

24.   This information comes from Web Designer Depot, "*The History and evolution of social media*", 7 October 2009, https://www.webdesignerdepot.com/2009/10/the-history-and-evolution-of-social-media/

25.   This information comes from forum-software.org, "*Forum Software Timeline 1994-2012*", https://www.forum-software.org/forum-software-timeline-from-1994-to-today

26.   Lene Bech, "*Is this the Web's first blog?*", Columbia Journalism Review, 5 November 2014, https://archives.cjr.org/behind_the_news/justin_hall_blog_web.php

27.   This information comes from Blogads, https://www.blogads.com/

28.   Paul Festa, "*Blogger founder leaves Google*", cnet, 5 October 2004, https://www.cnet.com/news/blogger-founder-leaves-google/

29.   This information comes from CBS News, https://www.cbsnews.com/pictures/then-and-now-a-history-of-social-networking-sites/7/

30.   This information comes from wpbegineer, https://www.wpbeginner.com/news/the-history-of-wordpress/

31.   Ace Exford, "*The History of Youtube*", engadget.com, 10 November 2016 https://www.engadget.com/2016-11-10-the-history-of-youtube.html

32.   Andrew Ross Sorkin and Jeremy W. Peters, "*Google to Acquire YouTube for $1.65 Billion*", The New York Times, 9 October 2006, https://www.nytimes.com/2006/10/09/business/09cnd-deal.html

33.   Mark Hall, "*Facebook*", Britannica, https://www.britannica.com/topic/Facebook

34.   This information comes from IGI Global,

https://www.igi-global.com/dictionary/i-found-myself-retweeting/30754

35.   Sherilynn Macale, "*A rundown of Reddit's history and community [Infographic]*", TNW, 14 October 2011, https://thenextweb.com/socialmedia/2011/10/14/a-rundown-of-reddits-history-and-community-infographic/

36.   Nicholas Carlson, "*INSIDE PINTEREST: An Overnight Success Four Years In The Making*", Business Insider, 2 May 2012, https://www.businessinsider.com.au/inside-pinterest-an-overnight-success-four-years-in-the-making-2012-4?r=US&IR=T

37.   Raisa Bruner, "*A Brief History of Instagram's Fateful First Day*", TIME, 16 July 2016, https://time.com/4408374/instagram-anniversary/

38.   Joe McFerrin, "*The History of eCommerce: How Did it All Begin?*", iwd, https://www.iwdagency.com/blogs/news/the-history-of-ecommerce-how-did-it-all-begin

39.   This information comes from History of Information, "*Book Stacks Unlimited Was a Precursor to Amazon.com's Online Bookstore*", https://www.historyofinformation.com/detail.php?id=3722

40.   This information comes from ebay, "*Our History*",https://www.ebayinc.com/company/our-history/

41.   Avery Hartmans, "*'Amazon' wasn't the original name of Jeff Bezos' company, and 14 other little-known facts about the early days of Amazon*", Business Insider, 17 July 2020, https://www.businessinsider.com/jeff-bezos-amazon-history-facts-2017-4?r=AU&IR=T

42.   Brian o' Connell, "*History of Paypal: Timeline and Facts*", The Street, 26 August 2019, https://www.thestreet.com/technology/history-of-paypal-15062744

43.   This information comes from Marketplace Pulse, "*Alibaba*", https://www.marketplacepulse.com/alibaba

44.   This information comes from BBC, "*Amazon becomes world's most valuable public company*", 8 January 2019, https://www.bbc.com/news/business-46793466

45.   This information comes from Alibaba.com, "*Alibaba.com Launches AliExpress to Bring Industry Leading Security, Choice, Flexibility, Convenience and Profitability to Small Wholesalers and Retailers*", 25 April 2010, https://news.alibaba.

com/article/detail/alibaba/100285058-1-alibaba.com-launches-aliexpress-bring-industry.
html

46.   This information comes from Compare.com, https://www.compare.com/ways-to-save/gig-economy/upwork-guide

47.   This information comes from steemit, "*History of Upwork*", https://steemit.com/history/@mdhasib11/history-of-upwork

48.   This information comes from fastcompany, "*How Fiverr's Founders Are Creating An Online Marketplace For Freelancers*", 17 September 2014, https://www.fastcompany.com/3035725/how-fiverrs-founders-are-creating-an-online-marketplace-for-freelance

49.   This information comes from Xcellab Magazine, "*Peer to peer Network Explained*", 17 December 2019, https://medium.com/xcellab-magazine/peer-to-peer-network-explained-c5038f6e8366

50.   Tom Lamont, "*Napster: the day the music was set free*", The Guardian, 24 February 2013, https://www.theguardian.com/music/2013/feb/24/napster-music-free-file-sharing

51.   Whitson Gordon, "*How to Use BitTorrent*", PC Magazine, 12 September 2019, https://au.pcmag.com/how-to/63432/how-to-use-bittorrent

52.   Dan Blystone, "*The Story of Uber*", Investopedia, 25 June 2019, https://www.investopedia.com/articles/personal-finance/111015/story-uber.asp

53.   Satoshi Nakamoto, "*Bitcoin: A Peer-to-Peer Electronic Cash System*", bitcoin.org,

54.   This information comes from Raspberrypi.org, https://www.raspberrypi.org/

55.   Chataut R, Akl R. Massive MIMO Systems for 5G and Beyond Networks-Overview, Recent Trends, Challenges, and Future Research Direction. Sensors (Basel). 2020;20(10):2753. Published 2020 May 12. doi:10.3390/s20102753

56.   Amy Nordrum, Kristen Clark and IEEE Spectrum Staff, "*Everything You Need to Know About 5G*", IEEE Spectrum, 27 January 2017, https://spectrum.ieee.org/video/telecom/wireless/everything-you-need-to-know-about-5g

57.   Amy Nordrum, Kristen Clark and IEEE Spectrum Staff, "*5G Bytes: Beamforming Explained*", IEEE Spectrum, 15 July 2017, https://spectrum.ieee.org/video/telecom/wireless/5g-bytes-beamforming-explained

Chapter 8

1.   Martin Childs, "*John McCarthy: Computer scientist known as the father of AI*", Independent, 1 November 2011, https://www.independent.co.uk/news/obituaries/john-mccarthy-computer-scientist-known-as-the-father-of-ai-6255307.html

2.   A. M. TURING, I.—Computing Machinery and Intelligence, Mind, Volume LIX, Issue 236, October 1950, Pages 433–460,

3.   Jazib Zaman, "*The AI Revolution: Has it happened yet?*", Techengage, 27 August 2020, https://techengage.com/the-ai-revolution-has-it-happened-yet/

4.   Harry Collins, "*Turing Test: why it still matters*", The Conversation, 3 October 2019, https://theconversation.com/turing-test-why-it-still-matters-123468

5.   This information comes from debategraph.org, "*PARRY*", https://debategraph.org/Details.aspx?nid=263

6.   Jeff Blagdon, "*AI with 13-year-old boy's personality wins top prize at world's biggest Turing test*", The Verge, 27 June 2012, https://www.theverge.com/2012/6/27/3120135/eugene-goostman-ukrainian-boy-ai-turing-test

7.   Sebastian Anthony, "*Facebook's facial recognition software is now as accurate as the human brain, but what now?*", ExtremeTech, 19 March 2014, https://www.extremetech.com/extreme/178777-facebooks-facial-recognition-software-is-now-as-accurate-as-the-human-brain-but-what-now

8.   Alex Hern, "*Real life CSI: Google's new AI system unscrambles pixelated faces*", ExtremeTech, 19 March 2014, https://www.theguardian.com/technology/2017/feb/08/google-ai-system-pixelated-faces-csi

9.   Hein de Haan, "*Deep Blue and AlphaZero: Comparing Giants of Artificial Intelligence*", Medium, 18 June 2019, https://medium.com/datadriveninvestor/deep-blue-and-alphazero-comparing-giants-of-artificial-intelligence-2c57081dd762

10.	This information coms from aiva.ai, https://www.aiva.ai/

11.	Amber Healy, "*Meet AIVA, the world's first AI music composer*", Geeks & Beats, 31 January 2018, https://www.geeksandbeats.com/2018/01/meet_avia_worlds_first_ai_music_composer/

12.	J.M. Porup, "*How and why deepfake videos work — and what is at risk*", CSO Australia, 10 April 2019, https://www.csoonline.com/article/3293002/deepfake-videos-how-and-why-they-work.html

13.	Ilija Mihajlovic, "*Everything You Ever Wanted To Know About Computer Vision*", Towards Data Science, 26 April 2019, https://towardsdatascience.com/everything-you-ever-wanted-to-know-about-computer-vision-heres-a-look-why-it-s-so-awesome-e8a58dfb641e?gi=c66309eb49a8

14.	Rohan Gupta, "*Breaking Down Facial Recognition: The Viola-Jones Algorithm*", Data Science, 7 August 2019, https://towardsdatascience.com/the-intuition-behind-facial-detection-the-viola-jones-algorithm-29d9106b6999?gi=c4bfe9b664e6

15.	Chrissy Kidd, "*NLU vs NLP: What's the difference?*", bmc.com, 28 May 2018, https://www.bmc.com/blogs/nlu-vs-nlp-natural-language-understanding-processing/

16.	Mitusha Arya, "*A brief history of Chatbots*", chatbotslife.com, 11 March 2019, https://chatbotslife.com/a-brief-history-of-chatbots-d5a8689cf52f?gi=694f65add62d

17.	Alex Debecker, "*A Closer Look at Chatbot Alice*", ubisend, 4 May 2017, https://blog.ubisend.com/discover-chatbots/chatbot-alice

18.	Andrew Griffin, "*Factbook's artificial intelligence robots shut down after they start talking to each other in their own language*", independent, 31 July 2017, https://www.independent.co.uk/life-style/facebook-artificial-intelligence-ai-chatbot-new-language-research-openai-google-a7869706.html

19.	Sam Wong, "*Google Translate AI invents its own language to translate with*", News Scientist, 30 November 2016, https://www.newscientist.com/article/2114748-google-translate-ai-invents-its-own-language-to-translate-with/

20.	Caroline Perry, "*The 1000 robot swarm*", The Harvard Gazette, 14 August

2014, https://news.harvard.edu/gazette/story/2014/08/the-1000-robot-swarm/

21.   Jonathan Amos, *"Termites inspire robot builders"*, BBC, 13 February 2014, https://www.bbc.com/news/science-environment-26025566

22.   David Delony, *"What is the 'AI winter' and how did it affect AI research?"*, Techopedia, https://www.techopedia.com/what-is-the-ai-winter-and-how-did-it-affect-ai-research/7/33404

23.   Naveen Joshi, *"How Far Are We From Achieving Artificial General Intelligence?"*, Forbes, 10 June 2019, https://www.forbes.com/sites/cognitiveworld/2019/06/10/how-far-are-we-from-achieving-artificial-general-intelligence/#348fb22a6dc4

Chapter 9

1.   Ehud Shapiro, *"Doctor in a cell"*, European Research Council, https://erc.europa.eu/projects-figures/stories/doctor-cell

2.   Tan SY, Tatsumura Y. Alexander Fleming (1881-1955): Discoverer of penicillin. Singapore Med J. 2015;56(7):366-367

3.   The information comes from the World Health Organization, *"Antibiotic resistance"*, https://www.who.int/news-room/fact-sheets/detail/antibiotic-resistance

4.   Hengbo Zhu, Li Wei, Ping Niu, *"The Novel coronavirus outbreak in Wuhan, China"*, Research Gate, 02 March 2020, https://www.researchgate.net/publication/339628619_The_novel_coronavirus_outbreak_in_Wuhan_China

5.   Lundstrom K. Viral Vectors in Gene Therapy. Diseases. 2018;6(2):42. Published 2018 May 21.

6.   Andreas Junk and Falk Riess, *"From an idea to a vision: There's plenty of room at the bottom"*, American Journal of Physics, 22 August 2006, https://aapt.scitation.org/doi/10.1119/1.2213634

7.   K. Eric. Drexler, *"Engines of Creation: The coming era of nanotechnology"*, Anchor Books, Doubleday, 1986, http://web.mit.edu/cortiz/www/3.052/3.052CourseReader/3_EnginesofCreation.pdf

8.   Colin Jeffrey, "*World's smallest engine powered by a single atom*", NewAtlas, 18 April 2016, https://newatlas.com/quantum-single-atom-engine/42849/

9.   Alex Marras, Lifeng Zhou, collaboration with Dr. Haijun Su,"*DNA Origami Machines and Mechanisms (DOMM)*", The Ohio State University, https://nbl.osu.edu/ DOMM

10.   Vaughan, O. Folded into three dimensions. Nature, Nanotech (2009). https://www.nature.com/articles/nnano.2009.145

11.   Phillip Broadwith, "*Nano-boxes from DNA origami*", Chemistry World, 7 May 2009, https://www.chemistryworld.com/news/nano-boxes-from-dna-origami-/3002630. article

12.   Edd Gent, "*4 Ways Scientists Hope Nanobots Will Make You Healthier*", Singularity Hub, 7 March 2017, https://singularityhub.com/2017/03/07/4-ways-scientists-hope-nanobots-will-make-you-healthier/

13.   Michael Berger, "*Nanotechnology machines' interaction with living systems*", nano werk, 05 July 2018, https://www.nanowerk.com/spotlight/spotid=50597.php

14.   This information comes from Medline Plus, "*What are genome editing and CRISPR-Cas9?*", https://medlineplus.gov/genetics/understanding/genomicresearch/ genomeediting/

15.   The information comes from The Harvard University, "*CRISPR: A game-changing genetic engineering technique*", 31 July 2014, http://sitn.hms.harvard.edu/ flash/2014/crispr-a-game-changing-genetic-engineering-technique/

16.   Meilin Zhu, "*CRISPR Acquired Resistance Against Viruses (2007)*", The Embryo Project Encyclopedia, https://embryo.asu.edu/pages/crispr-acquired-resistance-against-viruses-2007

17.   This information comes from cancer.org, "*How Chemotherapy Drugs Work*", https://www.cancer.org/treatment/treatments-and-side-effects/treatment-types/ chemotherapy/how-chemotherapy-drugs-work.html

18.   This information comes from cancer.gov, "*CAR T Cells: Engineering Patients' Immune Cells to Treat Their Cancers*",

https://www.cancer.gov/about-cancer/treatment/research/car-t-cells

19.   Denise Grady, "*In Girl's last hope, Altered Immune Cells Beat leukemia*", The New York Times, 9 December 2012, https://www.nytimes.com/2012/12/10/health/a-breakthrough-against-leukemia-using-altered-t-cells.html

20.   Dr. Francis Collins, "*FDA Approves First CAR-T Cell Therapy for Pediatric Acute Lymphoblastic Leukemia*", directorsblog.nih.gov, 30 August 2017, https://directorsblog.nih.gov/2017/08/30/fda-approves-first-car-t-cell-therapy-for-pediatric-acute-lymphoblastic-leukemia/

21.   Incorvaia L, Fanale D, Badalamenti G, et al. Programmed Death Ligand 1 (PD-L1) as a Predictive Biomarker for Pembrolizumab Therapy in Patients with Advanced Non-Small-Cell Lung Cancer (NSCLC). Adv Ther. 2019;36(10):2600-2617.

22.   Gonzalez H, Hagerling C, Werb Z. Roles of the immune system in cancer: from tumor initiation to metastatic progression. Genes Dev. 2018;32(19-20):1267-1284. doi:10.1101/gad.314617.118

23.   Wang K, Wei G, Liu D. CD19: a biomarker for B cell development, lymphoma diagnosis and therapy. Exp Hematol Oncol. 2012;1(1):36. Published 2012 Nov 29.

24.   Sopit Phetsang, "*Investigation of plasmonic gold nanostructure for enhancement of organic solar cells*", Niigata University, https://pubs.rsc.org/en/content/articlelanding/2019/na/c8na00119g#!divAbstract

25.   Estelrich J, Escribano E, Queralt J, Busquets MA. Iron oxide nanoparticles for magnetically-guided and magnetically-responsive drug delivery. Int J Mol Sci. 2015;16(4):8070-8101. Published 2015 Apr 10. doi:10.3390/ijms16048070

26.   Baylis F, McLeod M. First-in-human Phase 1 CRISPR Gene Editing Cancer Trials: Are We Ready?. Curr Gene Ther. 2017;17(4):309-319. doi:10.2174/1566523217666171121165935

27.   David Cyranoski, "*Chinese Scientists to Pioneer First Human CRISPR Trial*", Nature magazine, 22 July 2016, https://www.scientificamerican.com/article/chinese-scientists-to-pioneer-first-human-crispr-trial/

28. Thomas Sullivan, *"A Tough Road: Cost To Develop One New Drug Is $2.6 Billion; Approval Rate for Drugs Entering Clinical Development is Less Than 12%"*, Policy & Medicine, 21 March 2019, https://www.policymed.com/2014/12/a-tough-road-cost-to-develop-one-new-drug-is-26-billion-approval-rate-for-drugs-entering-clinical-de.html

29. This information comes from the National Human Genome Research Institute, *"The Human Genome Project"*, https://www.genome.gov/human-genome-project

30. This information comes from MGI, *"MGI's "life science super computer"* DNBSEQ-T7 officially delivered to business partners", https://en.mgitech.cn/News/info/id/10

31. Kimberly Powell, *"AI, Accelerated Computing Drive Shift to Personalized Healthcare"*, NVDIA, 17 December 2019, https://blogs.nvidia.com/blog/2019/12/17/ai-personalized-healthcare/

32. This information comes from Wyss Institute, https://wyss.harvard.edu/technology/human-organs-on-chips/

33. Bob Woods, *"It sounds futuristic, but it's not sci-fi: Human organs-on-a-chip"*, CNBC, 15 August 2017, https://www.cnbc.com/2017/08/14/fda-tests-groundbreaking-human-organs-on-a-chip.html

34. Suzanne Elvidge, *"Emulate raises $36M to develop organ-on-chip tech"*, Biopharma Dive, 21 June 2018, https://www.biopharmadive.com/news/emulate-raises-36m-to-develop-organ-on-chip-tech/526212/

35. This information comes from Bowhead Health, https://bowheadhealth.com/

36. Adam Conner-Simons and Rachel Gordon, *"Using AI to predict breast cancer and personalize care"*, MIT News, 7 May 2019, https://news.mit.edu/2019/using-ai-predict-breast-cancer-and-personalize-care-0507

37. Pranav Rajpurkar*, Jeremy Irvin*, Kaylie Zhu, Brandon Yang, Hershel Mehta, Tony Duan, Daisy Ding, Aarti Bagul, Curtis Langlotz, Katie Shpanskaya, Matthew P. Lungren, Andrew Y. Ng, "CheXNet: Radiologist-Level Pneumonia Detection on Chest X-Rays with Deep Learning", Standard ML Group,

https://stanfordmlgroup.github.io/projects/chexnet/

38.   This information comes from neuralink.com, https://neuralink.com/

39.   Natashah Hitti, *"Elon Musk's Neuralink implant will "merge" humans with AI"*, de zeen, 22 July 2019, https://www.dezeen.com/2019/07/22/elon-musk-neuralink-implant-ai-technology/

40.   Robert Gaunt and Jennifer Collinger, *"Brain-machine interfaces are getting better and better – and Neuralink's new brain implant pushes the pace"*, The Conversation, 19 July 2019, https://theconversation.com/brain-machine-interfaces-are-getting-better-and-better-and-neuralinks-new-brain-implant-pushes-the-pace-120562

41.   Arnav Lahiry, *"Will Elon Musk's Neuralink Shape The Future Of Humanity?"*, Oxford Student, 19 May 2020, https://www.oxfordstudent.com/2020/05/19/will-elon-musks-neuralink-shape-the-future-of-humanity/

42.   Elizabeth Lopatto, *"Elon Musk unveils Neuralink's plans for brain-reading 'threads' and a robot to insert them"*, The Verge, 16 July 2019, https://www.theverge.com/2019/7/16/20697123/elon-musk-neuralink-brain-reading-thread-robot

43.   Tim Urban, *"Neuralink and the Brain's Magical Future"*, Wait Buy Why, 20 April 2017, https://waitbutwhy.com/2017/04/neuralink.html

44.   Max Roser, Esteban Ortiz-Ospina and Hannah Ritchie, *"Life Expectancy"*, Our World in Data, First published in 2013; last revised in October 2019, https://ourworldindata.org/life-expectancy

45.   van Deursen JM. The role of senescent cells in ageing. Nature. 2014;509(7501):439-446. doi:10.1038/nature13193

46.   Petralia RS, Mattson MP, Yao PJ. Aging and longevity in the simplest animals and the quest for immortality. Ageing Res Rev. 2014;16:66-82. doi:10.1016/j.arr.2014.05.003

47.   This information comes from National Geographic, *"Galápagos Tortoise"*, https://www.nationalgeographic.com/animals/reptiles/g/galapagos-tortoise/

48.   David Derbyshine, *"Do lobsters hold the key to eternal life? Forget gastronomic indulgence, the crustacean can defy the ageing process"*, Daily Mail, 12

September 2013, https://www.dailymail.co.uk/sciencetech/article-2418252/Do-lobsters-hold-key-eternal-life-Forget-gastronomic-indulgence-crustacean-defy-ageing-process.html

49. John Roach, "*405-Year-Old Clam Called Longest-Lived Animal*", National Geographic, 29 October 2007, https://www.nationalgeographic.com/animals/2007/10/405-year-old-clam-called-longest-lived-animal/

50. Zane Bartlett, "*The Hayflick Limit*", The Embryo Project Encyclopedia, 14 November 2014, https://embryo.asu.edu/pages/hayflick-limit

51. This information comes from the Children's Hospital of Philadelphia, "*Vaccine Ingredients – Fetal Tissues*", https://www.chop.edu/centers-programs/vaccine-education-center/vaccine-ingredients/fetal-tissues

52. Boyce Rensberger, "*Microbes are immortal, so why aren't humans?*", Washington Post, 10 June 1998, https://www.washingtonpost.com/archive/1998/06/10/microbes-are-immortal-so-why-arent-humans/f7a95c80-e249-4000-81c0-de0ca34da2df/

53. This information comes from yourgenome.org, "*What is a telomere?*", 25 January 2016, https://www.yourgenome.org/facts/what-is-a-telomere

54. Jafri MA, Ansari SA, Alqahtani MH, Shay JW. Roles of telomeres and telomerase in cancer, and advances in telomerase-targeted therapies. Genome Med. 2016;8(1):69. Published 2016 Jun 20. doi:10.1186/s13073-016-0324-x

55. Tissenbaum HA. Using C. elegans for aging research. Invertebr Reprod Dev. 2015;59(sup1):59-63. doi:10.1080/07924259.2014.940470

56. Kenyon C. The first long-lived mutants: discovery of the insulin/IGF-1 pathway for ageing. Philos Trans R Soc Lond B Biol Sci. 2011;366(1561):9-16. doi:10.1098/rstb.2010.0276

57. Tullet JM. DAF-16 target identification in C. elegans: past, present and future. Biogerontology. 2015;16(2):221-234. doi:10.1007/s10522-014-9527-y

58. Sun X, Chen WD, Wang YD. DAF-16/FOXO Transcription Factor in Aging and Longevity. Front Pharmacol. 2017;8:548. Published 2017 Aug 23. doi:10.3389/fphar.2017.00548

59.   Vitale G, Pellegrino G, Vollery M, Hofland LJ. ROLE of IGF-1 System in the Modulation of Longevity: Controversies and New Insights From a Centenarians' Perspective. Front Endocrinol (Lausanne). 2019;10:27. Published 2019 Feb 1. doi:10.3389/fendo.2019.00027

60.   Clare Wilson, "*Calorie restriction diet extends life of monkeys by years*", New Scientist, 17 January 2017, https://www.newscientist.com/article/2118224-calorie-restriction-diet-extends-life-of-monkeys-by-years/

61.   Kevin Warwick, "*Project Cyborg 1.0*", http://www.kevinwarwick.com/project-cyborg-1-0/

62.   This information comes from pri.org, "*Engineering Extra Senses: Technology and the Human Body*", 14 November 2012, https://www.pri.org/stories/2012-11-14/engineering-extra-senses-technology-and-human-body

63.   This information comes fro openworm.org, http://openworm.org/

64.   Marblestone AH, Zamft BM, Maguire YG, et al. Physical principles for scalable neural recording. Front Comput Neurosci. 2013;7:137. Published 2013 Oct 21. doi:10.3389/fncom.2013.00137

Chapter 10

1.   Sleator RD. The story of Mycoplasma mycoides JCVI-syn1.0: the forty million dollar microbe. Bioeng Bugs. 2010;1(4):229-230. doi:10.4161/bbug.1.4.12465

2.   This information comes from the National Human Genome Research Institute, "*1859: Darwin Published On the Origin of Species, Proposing Continual Evolution of Species*", https://www.genome.gov/25520157/online-education-kit-1859-darwin-published-on-the-origin-of-species-proposing-continual-evolution-of-species

3.   The information comes from the U.S. National Library of Medicine, "*The Discovery of the Double Helix, 1951-1953*", https://profiles.nlm.nih.gov/spotlight/sc/feature/doublehelix

4.   This information comes from the National Human Genome Research Institute, "*June 2000 White House*",

https://www.genome.gov/10001356/june-2000-white-house-event

5.   Mindy Weisberger, "*Early Animal Life Exploded on Earth Even Earlier Than Once Thought*", Live Science, 13 March 2019, https://www.livescience.com/64977-rethinking-cambrian-explosion.html

6.   This information comes from Science Daily, "*Somatic cell nuclear transfer*", https://www.sciencedaily.com/terms/somatic_cell_nuclear_transfer.htm

7.   This information comes from the AnimalResearch.Info, "*Cloning Dolly the sheep*", 3 November 2014, http://www.animalresearch.info/en/medical-advances/timeline/cloning-dolly-the-sheep/

8.   Will Knight, "*Dolly the sheep dies young*", New Scientist, 14 February 2003, https://www.newscientist.com/article/dn3393-dolly-the-sheep-dies-young/

9.   Lagutina I, Fulka H, Lazzari G, Galli C. Interspecies somatic cell nuclear transfer: advancements and problems. Cell Reprogram. 2013;15(5):374-384. doi:10.1089/cell.2013.0036

10.   Gupta MK, Das ZC, Heo YT, et al. Transgenic chicken, mice, cattle, and pig embryos by somatic cell nuclear transfer into pig oocytes. Cell Reprogram. 2013;15(4):322-328. doi:10.1089/cell.2012.0074

11.   Paul Rincon, "*Fresh effort to cline extinct animal*", BBC, 22 November 2013, https://www.bbc.com/news/science-environment-25052233

12.   Mario L Major, "*This Extinct Animal Was Resurrected by Cloning, Only to Go Extinct Once Again*", Interesting Engineering, 29 November 2017, https://interestingengineering.com/this-extinct-animal-was-resurrected-by-cloning-only-to-go-extinct-once-again

13.   Zuo Y, Gao Y, Su G, Bai C, Wei Z, Liu K, Li Q, Bou S, Li G. Irregular transcriptome reprogramming probably causes thec developmental failure of embryos produced by interspecies somatic cell nuclear transfer between the Przewalski's gazelle and the bovine. BMC Genomics. 2014 Dec 16;15(1):1113. doi: 10.1186/1471-2164-15-1113. PMID: 25511933; PMCID: PMC4378013.

14.   Peter Dockrill, "*Scientists Have 'Revived' Cell Parts From a 28,000-Year-Old Extinct Woolly Mammoth*", Nature, 13 March 2019, https://www.sciencealert.com/scientists-have-reawakened-nuclei-from-an-extinct-mammoth-who-died-28-000-years-ago

15.   Hannah Devlin, "*Woolly mammoth on verge of resurrection, scientists reveal*", The Guardian, 17 February 2017, https://www.theguardian.com/science/2017/feb/16/woolly-mammoth-resurrection-scientists

16.   National Research Council (US). Sharing Laboratory Resources: Genetically Altered Mice: Summary of a Workshop Held at the National Academy of Sciences, March 23-24, 1993. Washington (DC): National Academies Press (US); 1994. 3, Genetically Altered Mice: A Revolutionary Research Resource. Available from: https://www.ncbi.nlm.nih.gov/books/NBK231336/

17.   Jon Van, "*In Oil-spill clean-ups, major tool off-limits*", chicagotribune, 18 June 1989, https://www.chicagotribune.com/news/ct-xpm-1989-06-18-8902100449-story.html

18.   Bruening G, Lyons J. 2000. The case of the FLAVR SAVR tomato. Calif Agr 54(4):6-7.

19.   Ian Sample, "*Anti-malarial mosquitoes' created using controversial genetic technology*", The Guardian, 24 November 2015, https://www.theguardian.com/science/2015/nov/23/anti-malarial-mosquitoes-created-using-controversial-genetic-technology

20.   Schmidt M. Xenobiology: a new form of life as the ultimate biosafety tool. Bioessays. 2010;32(4):322-331. doi:10.1002/bies.200900147

21.   The information comes from biobricks.org, https://biobricks.org/

22.   Robert F. Service, "*Synthetic microbe lives with fewer than 500 genes*", Science Mag, Science Mag, 24 March 2016, https://www.sciencemag.org/news/2016/03/synthetic-microbe-lives-fewer-500-genes

23.   Sleator RD. JCVI-syn3.0 - A synthetic genome stripped bare!. Bioengineered. 2016;7(2):53-56. doi:10.1080/21655979.2016.1175847

24.   This information comes from the Skeptical Chemist, "*Xeno Nucleic Acids –
Research into Modified DNA*", published 21 August 2020, https://theskepticalchemist.
com/xna-research-synthetic-nucleic-acids/

25.   Dien VT, Morris SE, Karadeema RJ, Romesberg FE. Expansion of the genetic
code via expansion of the genetic alphabet. Curr Opin Chem Biol. 2018 Oct;46:196-
202. doi: 10.1016/j.cbpa.2018.08.009. Epub 2018 Sep 8. PMID: 30205312; PMCID:
PMC6361380.

26.   The information comes from ACS Chemistry for Life,
https://www.acs.org/content/acs/en/pressroom/presspacs/2015/acs-presspac-
may-27-2015/expanding-the-code-of-life-with-new-letters.html

27.   This information comes from Scripps Research, "*Scientists Create First Semi-
Synthetic Organism that Stores and Retrieves Unnatural Information*", 29 November
2017, https://www.scripps.edu/news-and-events/press-room/2017/20171130romesberg.
html

28.   Oliver MJ. Why we need GMO crops in agriculture. Mo Med.
2014;111(6):492-507.

29.   This information comes from genengnews.com, "*Artificial Proteins Give
Living Cells a Computational Upgrade*", 6 April 2020, https://www.genengnews.com/
news/artificial-proteins-give-living-cells-a-computational-upgrade/

30.   Anosova I, Kowal EA, Dunn MR, Chaput JC, Van Horn WD, Egli M. The
structural diversity of artificial genetic polymers. Nucleic Acids Res. 2016;44(3):1007-
1021. doi:10.1093/nar/gkv1472

31.   Michael Marshall, "*The secret of how life on Earth began*", BBC, 31 October
2016, http://www.bbc.com/earth/story/20161026-the-secret-of-how-life-on-earth-began

32.   Benjamin Radford, "*Hybrid creatures: The Real Science of 'Splice'*", Live
Science, 2 June 2010, https://www.livescience.com/10659-hybrid-creatures-real-science-
splice.html

33.   The information comes from BBC, "*The goats with spider genes and silk in
their milk*", BBC, https://www.bbc.com/news/av/science-environment-16554357

Chapter 11

1.	This information comes from nano.gov, "*What's So Special about the Nanoscale?*", https://www.nano.gov/nanotech-101/special

2.	This information comes from the University of Manchester, "*Discovery of graphene*", https://www.graphene.manchester.ac.uk/learn/discovery-of-graphene/

3.	University of Central Florida. "*Graphene transistor could mean computers that are 1,000 times faster: Next-gen, carbon-based transistors would far outperform today's silicon ones.*" ScienceDaily. ScienceDaily, 13 June 2017.

4.	Michael Irving, "*Ultra-low power graphene-based transistor could enable 100 GHz clock speeds*", New Atlas, 25 May 2016, https://newatlas.com/graphene-transistor-clock-speeds-100-ghz/43499/

5.	This information comes from first grapheme, "*New Graphene Supercapacitor Materials Offer Fast Charging for Electric Vehicles*", 15 May 2020, https://firstgraphene.net/new-graphene-supercapacitor-materials-offer-fast-charging-for-electric-vehicles/

6.	Tonelli FM, Goulart VA, Gomes KN, Ladeira MS, Santos AK, Lorençon E, Ladeira LO, Resende RR. Graphene-based nanomaterials: biological and medical applications and toxicity. Nanomedicine (Lond). 2015;10(15):2423-50. doi: 10.2217/nnm.15.65. Epub 2015 Aug 5. PMID: 26244905.

7.	This information comes from humanityplus.wordpress.com, "*Graphene*", 13 October 2017, https://humanityplus.wordpress.com/2017/10/13/graphene/

8.	University of Arkansas. "*Physicists build circuit that generates clean, limitless power from graphene: Researchers harnessed the atomic motion of graphene to generate an electrical current that could lead to a chip to replace batteries.*" ScienceDaily. ScienceDaily, 2 October 2020.

9.	This information comes from the university of Arkansas, "*Tiny but Mighty: Revolutionary discovery harvests limitless power of graphene.*", https://www.uark.edu/determined/features/tiny-but-mighty/index.php

10.   Alyn Griffiths, "*Graphene-based water filter produces drinkable water in just one step*", de zeen, 4 March 2018, https://www.dezeen.com/2018/03/04/graphene-water-filter-produces-drinkable-water-in-just-one-step/

11.   Mervyn Piesse, "*Research and Development in the Global Desalination Industry*", Future Directions, 12 April 2018, https://www.futuredirections.org.au/publication/research-development-global-desalination-industry/

12.   This information comes from Technology Review, "*Radha Boya*", https://www.technologyreview.com/innovator/radha-boya/

13.   Isabelle Dubach, "*Graphene-iron filters a promising gas separation tool: research*", UNSW, 27 March 2020, https://newsroom.unsw.edu.au/news/science-tech/graphene-iron-filters-promising-gas-separation-tool-research

14.   Rice University. "*Graphene oxide soaks up radioactive waste: U.S., Russian researchers collaborate on solution to toxic groundwater woes.*" ScienceDaily. ScienceDaily, 8 January 2013.

15.   Wang, F., Li, H., Liu, Q. et al. A graphene oxide/amidoxime hydrogel for enhanced uranium capture. Sci Rep 6, 19367 (2016). https://doi.org/10.1038/srep19367

16.   Susan Bauer, "*Seawater yields first grams of yellowcake*", Pacific Northwest, 13 June 2018, https://www.pnnl.gov/news/release.aspx?id=4514

17.   Yuehuan Wei, Jianyu Long, Francesco Lombardi, Zhiheng Jiang, Jingqiang Ye, Kaixuan Ni, "*Development of a Sealed Liquid Xenon Time Projection Chamber with a Graphene-Coated Electrode*", Cornell University, 31 July 2020, arXiv:2007.16194

18.   Jessica Zeitz, "*A New Fibre Developed by Tsinghua University in Beijing Could be Strong Enough to Build a Space Elevator*", Asgardia, 6 November 2018, https://asgardia.space/en/news/asgardia-space-news-a-new-fibre-developed-by-tsinghua-university-in-beijing-could-be-strong-enough-to-build-a-space-elevator

19.   This information comes from Slashdot, "*China Produces Nano Fibre That Can Lift 160 Elephants - and a Space Elevator*", https://science.slashdot.org/story/18/10/28/0334249/china-produces-nano-fibre-that-can-lift-160-elephants---and-a-space-elevator

20.   Vassili Fedotov, "Metamaterials", Springerlink, https://link.springer.com/chapter/10.1007/978-3-319-48933-9_56

Chapter 12

1.   Andrew J. Hawkins, *"How Tesla changed the auto industry forever"*, the verge, 28 July 2017, https://www.theverge.com/2017/7/28/16059954/tesla-model-3-2017-auto-industry-influence-elon-musk

2.   Kirsten Korosec, *"Tesla plans to launch a robotaxi network in 2020"*, Techcrunch, 23 April 2019, https://techcrunch.com/2019/04/22/tesla-plans-to-launch-a-robotaxi-network-in-2020/

3.   This information comes from waymo, https://waymo.com/

4.   Kyle Wiggers, *"Neolix raises $29 million to mass-produce autonomous delivery shuttles"*, 11 March 2020, https://venturebeat.com/2020/03/11/neolix-raises-29-million-to-mass-produce-autonomous-delivery-shuttles/

5.   Steve Ranger, *"What is Hyperloop? Everything you need to know about the race for super-fast travel"*, 16 August, 2019, https://www.zdnet.com/article/what-is-hyperloop-everything-you-need-to-know-about-the-future-of-transport/

6.   Lucas Asher, *"Op-ed: The hyperloop will revolutionize transportation in the post-coronavirus world"*, CNBC, 2 September 2020, https://www.cnbc.com/2020/09/02/hyperloop-will-revolutionize-transportation-in-post-coronavirus-world.html

7.   Steve Ranger, *"What is Hyperloop? Everything you need to know about the race for super-fast travel"*, 16 August, 2019, https://www.zdnet.com/article/what-is-hyperloop-everything-you-need-to-know-about-the-future-of-transport/

8.   Tom Page, *"Hyperloop for cargo aims to deliver at over 600 mph"*, CNN, 4 May 2018, https://edition.cnn.com/2018/05/04/tech/hyperloop-dp-world-cargospeed-announcement/index.html

9.   Francesca Street, *"How long until Hyperloop is here?"*, CNN, 11 December 2019, https://edition.cnn.com/travel/article/how-long-hyperloop/index.html

10. Nadia Drake, "*Elon Musk: In Seven Years, SpaceX Could Land Humans on Mars*", National Geographic, 29 September 2017, https://www.nationalgeographic.com/news/2017/09/elon-musk-spacex-mars-moon-bfr-rockets-space-science/

11. Sean O'Kane, "*Elon Musk proposes city-to-city travel by rocket, right here on Earth*", the verge, 29 September 2017, https://www.theverge.com/2017/9/29/16383048/elon-musk-spacex-rocket-transport-earth-travel

12. Brett Williams, "*Dubai's self-flying taxis are primed for takeoff later this year*", Mashable, 20 June 2017, https://mashable.com/2017/06/19/dubai-autonomous-taxi-service-scheduled/

13. The information comes from Uber Elevate, https://www.uber.com/us/en/elevate/

14. Lauren M. Johnson, "*A man flying a jetpack was reported by pilots above Los Angeles*", CNN, 2 September 2020, https://edition.cnn.com/2020/09/01/us/jetpack-lax-trnd/index.html

15. This information comes from KnowledgeNuts, "*The World's First Jetpack Was Built In The 1950s*", 8 September 2013, https://knowledgenuts.com/2013/09/08/the-worlds-first-jetpack-was-built-in-the-1950s/

16. This information comes from DXC.technology, "*Australian First: U.K. Inventor Richard Browning's Human Flight Demonstration*", https://www.dxc.technology/au/flxwd/143870-dxc_technology_brings_real_life_iron_man_richard_browning_to_australia_for_the_first_time

Chapter 13

1. The information comes from i4.0today, "*The digital revolution – An insight into Industry 4.0*", http://i40today.com/why-industry-4-0/

2. Richard D'Aveni, "*The 3-D Printing Revolution*", Harvard Business Review, May 2015, https://hbr.org/2015/05/the-3-d-printing-revolution

3. This information comes from Smart Energy International, "*Lighting a foothold into IoT deployments for cities*", 11 October 2017,

https://www.smart-energy.com/regional-news/north-america/smart-street-lighting/

4.   Umow Lai, *"Automated Waste Collection in a Smart City"*, umowlai, 29
August 2019, http://umowlai.com.au/automated-waste-collection-in-a-smart-city/

5.   Ai Lei Tao, *"Singapore models entire country in 3D with smart map"*,
Computer Weekly, 24 August 2016, https://www.computerweekly.com/news/450302992/
Singapore-models-entire-country-in-3D-with-smart-map

Chapter 14

1.   This information comes from www.w3.org, *"Tim Berners-Lee"*, https://www.
w3.org/People/Berners-Lee/

2.   This information comes from CERN, *"About CERN"*, https://home.cern/
node/5011

3.   This information comes from CERN, *"The Large Hadron Collider"*,
https://home.cern/science/accelerators/large-hadron-collider

4.   This information comes from CERN, *"First beam in the LHC – accelerating
science"*, https://home.cern/news/press-release/cern/first-beam-lhc-accelerating-science

5.   This information comes from open.edu, *"W and Z bosons"*, https://www.open.
edu/openlearn/science-maths-technology/particle-physics/content-section-8.1

6.   Brian Greene, *"How he Higgs Boson was Found"*, smithsonianmag, July 2013,
https://www.smithsonianmag.com/science-nature/how-the-higgs-boson-was-
found-4723520/

7.   This information comes from CERN, *"The Standard Model"*, https://home.
cern/science/physics/standard-model

8.   This information comes from ABC News, *"CERN plans new particle
accelerator four times bigger than Large Hadron Collider"*, 16 January 2019,
https://www.abc.net.au/news/2019-01-16/cern-plans-new-particle-accelerator-four-times-
bigger-than-lhc/10718874

Chapter 15

1.	This information comes from lincolnparkboe.org, "*Atomic Theory Timeline*", http://www.lincolnparkboe.org/userfiles/33/Classes/239/Atomic%20Theory%20 Information%20Book.pdf

2.	This information comes from nobelprize.prg, "*Erwin Schrödinger*" https://www.nobelprize.org/prizes/physics/1933/schrodinger/biographical/

3.	This information comes from Stanford Encyclopedia of Philosophy, "*Copenhagen Interpretation of Quantum Mechanics*", https://plato.stanford.edu/entries/ qm-copenhagen/

4.	A. Einstein, B. Podolsky, and N. Rosen, "*Can Quantum-Mechanical Description of Physical Reality Be Considered Complete?*" Phys. Rev. 47, 777 – Published 15 May 1935, doi.org/10.1103/PhysRev.47.777

5.	Gabriel Popkin, "*Einstein's 'spooky action at a distance' spotted in objects almost big enough to see*", Science Magazine, 25 April 2018, https://www.sciencemag. org/news/2018/04/einstein-s-spooky-action-distance-spotted-objects-almost-big-enough-see

6.	Stephen Boughn, "*Making Sense of Bell's Theorem and Quantum Nonlocality*",Springlink, 27 March 2017, https://link.springer.com/article/10.1007/ s10701-017-0083-6

7.	JR Minkel, "*Quantum Spookiness Spans the Canary Islands*", Scientific American, 9 March 2007,https://www.scientificamerican.com/article/entangled-photons-quantum-spookiness/

8.	Paul E. Black, D. Richard Kuhn, Carl J. Williams, "*Quantum Computing and Communication*", National Institute of Standards and technology, doi.org/10.1016/j. compeleceng.2013.10.008

Chapter 16

1. This information comes from Ohio-state.edu, *"Real-World Relativity: The GPS Navigation System"*, http://www.astronomy.ohio-state.edu/~pogge/Ast162/Unit5/gps.html

2. Ringbauer, M. et al. Experimental simulation of closed timelike curves. Nat. Commun. 5:4145 doi: 10.1038/ncomms5145 (2014).

3. Clara Moskowitz, *"Physicists Find a Link between Wormholes and Spooky Action at a Distance"*, Scientific American, 11 December 2013, https://www.scientificamerican.com/article/wormholes-quantum-entanglement-link/

Chapter 17

1. This information comes from nasa.gov, "Konstantin E. Tsiolkovsky", https://www.nasa.gov/audience/foreducators/rocketry/home/konstantin-tsiolkovsky.html

2. This information comes from nasa.gov, "The Tyranny of the Rocket Equation", https://www.nasa.gov/mission_pages/station/expeditions/expedition30/tryanny.html

3. This information comes from nasa.gov, "Dr. Robert H. Goddard, American Rocketry Pioneer", https://www.nasa.gov/centers/goddard/about/history/dr_goddard.html

4. This information comes from National Space Society," Reaching for the High Frontier Chapter 1", https://space.nss.org/reaching-for-the-high-frontier-chapter-1/

5. This information comes from nasa.gov," Goddard space flight center", https://www.nasa.gov/centers/goddard/news/goddard-features.html

6. This information comes from Britannica.com, "V-2 missile", https://www.britannica.com/technology/V-2-missile

7. Richard Stone, "'National pride is at stake.' Russia, China, United States race to build hypersonic weapons", Science Magazine, 8 January 2020, https://www.sciencemag.org/news/2020/01/national-pride-stake-russia-china-united-states-race-build-hypersonic-weapons

8. Owen Edward, "Wernher von Braun's V-2 Rocket", Smithsonian Magazine, August 2011, https://www.smithsonianmag.com/arts-culture/wernher-von-brauns-v-2-

rocket-12609128/

9.   This information comes from nasa.gov, "Sputnik and the Dawn of the Space Age", https://history.nasa.gov/sputnik.html

10.   This information comes from nasa.gov, "Explorer and Early Satellites", https://www.nasa.gov/mission_pages/explorer/explorer-overview.html

11.   This information comes from history.com, "NASA created", https://www.history.com/this-day-in-history/nasa-created

12.   This information comes from space.com, "May 25, 1961: JFK's Moon Shot Speech to Congress", https://www.space.com/11772-president-kennedy-historic-speech-moon-space.html

13.   This information comes from nasa.gov, "What Was the Gemini Program?", 16 March 2011, https://www.nasa.gov/audience/forstudents/5-8/features/nasa-knows/what-was-gemini-program-58.html

14.   This information comes from nasa.gov, "What was the Apollo Program?" https://www.nasa.gov/audience/forstudents/5-8/features/nasa-knows/what-was-apollo-program-58.html

15.   This information comes from nasa.gov, "What Was the Saturn V?" https://www.nasa.gov/audience/forstudents/5-8/features/nasa-knows/what-was-the-saturn-v-58.html

16.   This information comes from astronautix.com, "Korolev, Sergei Pavlovich", http://www.astronautix.com/k/korolev.html

17.   Jason Rodrigues, "The Challenger space shuttle disaster at 30: how the Guardian covered the tragedy", the Guardian, 29 January 2016, https://www.theguardian.com/science/from-the-archive-blog/2016/jan/28/space-shuttle-challenger-nasa-disaster-30-1986

18.   Elizabeth Howell, "Columbia Disaster: What Happened, What NASA Learned", space.com, 1 February 2019, https://www.space.com/19436-columbia-disaster.html

19.   Mike Wall, "President Obama's Space Legacy: Mars, Private Spaceflight and More", space.com, 20 January 2017,

https://www.space.com/35394-president-obama-spaceflight-exploration-legacy.html

20.   This information comes from nasa.gov, "Apollo-Soyuz: An Orbital Partnership Begins", 10 July 2015, https://www.nasa.gov/topics/history/features/astp.html

21.   This information comes from nasa.gov, "International Space Station", https://www.nasa.gov/mission_pages/station/cooperation/index.html

22.   Bill Birtles, "China's Mars mission is a glimpse into its space ambitions. But we're not in a space race yet", ABC news, 30 July 2020,

https://www.abc.net.au/news/2020-07-30/china-and-the-us-head-to-mars-marking-space-golden-era/12489972

23.   Callum Hoare, "China's space ban: NASA 'forbidden from sharing info with Beijing' over security fears", Express, 23 July 2020,

https://www.express.co.uk/news/world/1313337/china-space-ban-tianwen1-mars-rover-nasa-beijing-security-fears-iss-us-legislation-spt

24.   Carolyn Collins Petersen, "The History of the Chinese Space Program", ThoughtCo.,3 May 2018, https://www.thoughtco.com/chinese-space-program-4164018

25.   Adam Mann, "China's Chang'e Program: Missions to the Moon", space.com, 1 February 2019, https://www.space.com/43199-chang-e-program.html

26.   Elizabeth Howell, "Tiangong-1: China's First Space Station", space.com, 2 March 2018, https://www.space.com/27320-tiangong-1.html

27.   Matthew S. Schwartz, "China Becomes First Country To Land On Far Side Of Moon, State Media Announce", npr.org, 3 January 2019, https://www.npr.org/2019/01/03/681825762/china-becomes-first-country-to-land-on-far-side-of-moon-state-media-announces

28.   This information comes from nasa.gov, "NASA Reaches New Heights in 2015", https://www.nasa.gov/press-release/nasa-reaches-new-heights-in-2015/

29.   Kenneth Chang ,"Jeff Bezos Unveils Blue Origin's Vision for Space, and a Moon Lander", The NewYork Times, 9 May 2019, https://www.nytimes.com/2019/05/09/science/jeff-bezos-moon.html

30.   This information comes from Blue Origin, "New Shepard", https://www.blueorigin.com/new-shepard/

31.   This information comes from Blue Origin, "New Glenn", https://www.blueorigin.com/new-glenn/

32.   This information comes from Blue Origin, "Blue Engines", https://www.blueorigin.com/engines/

33.   Nick Statt, *"Jeff Bezos will sell Amazon stock every year to fund Blue Origin"*, the verge, 5 April 2017, https://www.theverge.com/2017/4/5/15200102/jeff-bezos-amazon-stock-blue-origin-space-travel-funding

34.   This information comes from Blue Origin, "Blue Moon", https://www.blueorigin.com/blue-moon/lunar-transport

35.   Alan Boyle, "Blue Origin and its partners deliver a lunar lander mock-up to NASA", Geekwire, 20 August 2020, https://www.geekwire.com/2020/blue-origin-partners-deliver-lunar-lander-mock-nasa-try/

36.   Elizabeth Howell, "SpaceX: Facts About Elon Musk's Private Spaceflight Company", space.com, 16 December 2019, https://www.space.com/18853-spacex.html

37.   Aaron Rowe, "SpaceX Did It — Falcon 1 Made it to Space", wired, 20 September 2008, https://www.wired.com/2008/09/space-x-did-it/

38.   Rachael Thomas, "Super heavy-lift launch vehicles: How does Falcon Heavy stack up?",florida today, 9 April 2019, https://www.floridatoday.com/story/tech/science/space/2019/04/09/worlds-most-powerful-rockets-saturn-delta-falcon-heavy-sls-new-glenn/3411587002/

39.   Michael Collett, "Falcon Heavy: Elon Musk is sending this Tesla into space, but that's not why this is important", ABC News, 6 February 2018, https://www.abc.net.au/news/2018-02-06/elon-musk-is-sending-a-tesla-into-space-via-

the-falcon-heavy/9393256

40.   Genelle Weule, "NASA and Elon Musk's SpaceX to send astronauts to the ISS in the first crewed mission from the US since the space shuttle", ABC News, 26 May 2020, https://www.abc.net.au/news/science/2020-05-26/nasa-spacex-elon-musk-astronauts-commercial-mission-to-space/12283176

41.   Paul Rincon, "Water ice 'detected on Moon's surface'", BBC News, 21 August 2018, https://www.bbc.com/news/science-environment-45251370

42.   Jake Parks, "Moon Village: Humanity's first step toward a lunar colony?", Astronomy, 31 May 2019, https://astronomy.com/news/2019/05/moon-village-humanitys-first-step-toward-a-lunar-colony

43.   Karl Tate, "How Living on Mars Could Challenge Colonists (Infographic)", space.com, 17 February 2015, https://www.space.com/27202-living-on-mars-conditions-infographic.html

44.   This information comes from nasa.gov, "Overview: In-Situ Resource Utilization", https://www.nasa.gov/isru/overview/

45.   Mike Wall, "Elon Musk Floats 'Nuke Mars' Idea Again (He Has T-Shirts)", space.com, 17 August 2019, https://www.space.com/elon-musk-nuke-mars-terraforming.html

Recommended Readings
To further enhance your world views

The Dollar Crisis (**Richard Duncan**)

Corruption of Capitalism (**Richard Duncan**)

The Creature of Jekyll Island (**G. Edward Griffin**)

The End of Growth: Adapting to Our New Economic Reality (**Richard Heinberg**)

Peak Everything: Waking up to the century of declines (**Richard Heinberg**)

Limit to Growth (**Dennis Meadows, Donella Meadows, Jørgen Randers, William W. Behrens III**)

Rise of the Robots: Technology and the Threat of a Jobless Future (**Martin Ford**)

The Future of Humanity (**Michio Kaku**)

The Singularity is Near (**Ray Kurzweil**)

About the Author

Marco Chu Kwan Ching
Author, Investor

I began my profession as an electrical engineer in TOSHIBA after graduating from **_The University of New South Wales_** (UNSW) in 2009. I had a small web design business for indexing restaurants. I was firmly sitting on the dream of most undergraduates- a job and a part time business. With the collapse of the global economy in 2008, I first noticed how the effects of the financial crisis unfolded. I experienced the accelerating inflation rapidly eroding our wealth. I witnessed the foreclosures of businesses, income polarization, the interventions of the Government monetary policies. Even with little life experience on these subjects, I know something is not right. The current financial system is developing cracks. This sparked my interest in studying monetary history and the global economy. I set out to research the answers myself.

What I found shocked me to my core. The root of all the problems lie within our philosophy of money. The definition of money is flawed. Currency is not money. The original idea of money being a container to store the value of our labor, time, ideas, and talents are replaced by debts. Money, rather than being a store of value, becomes a plan to transfer our wealth away from us. My mission is to educate as many people as possible about these findings, so they are armed with the right knowledge to protect themselves and their family from this corrupt monetary system. That's why I am willing to give up my time to work on the material that now appears in the **_Corruption of Real Money Series_**.

Thank you for Reading!

Thank you for reading this book! I know you could have picked from a dozens of books about this subject, but you took a chance with mine and I appreciate it.

Lastly, I just have one small request,

If you believe that this book is worth sharing, would you take a few seconds to let your friends know about it too? If you love my work, please feel free to leave a positive feedback on Amazon and Goodreads.

My contact:
https://www.facebook.com/marco.chu.10
https://www.goodreads.com/author/show/15944678.Marco_Chu_Kwan_Ching

Corruption of Real Money Facebook Page
https://www.facebook.com/CorruptionOfRealMoney/

Corruption of Real Money Twitter Page
https://twitter.com/CorruptionofMon

Goodreads Page
https://www.goodreads.com/book/show/55606868-crystal-balls-of-the-21st-century

Corruption of Real Money Website
http://www.corruptionofrealmoney.com

***

Book Series by Marco Chu Kwan Ching

1 **Corruption of Real Money (Monetary History and Global Economy)**
   http://www.corruptionofrealmoney.com

2 **Terrorlands (Children's Horror Fiction)**
   http://www.terrorlands.com/